Planning Your
Research
and How to Write It

Planning Your Research and How to Write It

Aziz Nather

*Department of Orthopaedics Surgery, Yong Loo Lin School of Medicine,
National University Health System, Singapore*

NEW JERSEY • LONDON • SINGAPORE • BEIJING • SHANGHAI • HONG KONG • TAIPEI • CHENNAI

Published by

World Scientific Publishing Co. Pte. Ltd.

5 Toh Tuck Link, Singapore 596224

USA office: 27 Warren Street, Suite 401-402, Hackensack, NJ 07601

UK office: 57 Shelton Street, Covent Garden, London WC2H 9HE

Library of Congress Cataloging-in-Publication Data
Planning your research and how to write it : a practical guide for residents / [edited by] Aziz Nather.
   p. ; cm.
   Includes bibliographical references and index.
   ISBN 978-9814651035 (hardcover : alk. paper) -- ISBN 978-9814651042 (pbk. : alk. paper) -- ISBN 978-9814651059 (electronic : alk. paper)
   I. Nather, Aziz, editor.
   [DNLM: 1. Research Design. 2. Biomedical Research. 3. Writing. W 20.5]
   R850
   610.72'4--dc23

2014044834

British Library Cataloguing-in-Publication Data
A catalogue record for this book is available from the British Library.

Copyright © 2016 by World Scientific Publishing Co. Pte. Ltd.

All rights reserved. This book, or parts thereof, may not be reproduced in any form or by any means, electronic or mechanical, including photocopying, recording or any information storage and retrieval system now known or to be invented, without written permission from the publisher.

For photocopying of material in this volume, please pay a copying fee through the Copyright Clearance Center, Inc., 222 Rosewood Drive, Danvers, MA 01923, USA. In this case permission to photocopy is not required from the publisher.

Typeset by Stallion Press
Email: enquiries@stallionpress.com

I would like to dedicate this book to my wife, Suraiya Rahman, my children, Sharnaz, Zameer and Azad especially to my two grandchildren, Samira Tan (10 yrs) and Aliyah Tan (7 yrs old).

Mira and Liyah's charms, dynamism and bustling activities helps to keep their 'dada' forever young and productive.

Contents

Preface xi
About the Editor xiii
List of Contributors xv

Section I: Introduction 1

Lessons from Research: A Personal Experience
Aziz Nather 3

Section II: Planning Your Research 13

Chapter 1 Planning Research 15
*Aziz Nather, Jamie Xiang Lee Kee &
Haitong Mao*

Chapter 2 Procuring Research Grants 31
Haitong Mao & Aziz Nather

Chapter 3 Types of Research: An Overview 57
*Jamie Xiang Lee Kee, Haitong Mao &
Aziz Nather*

Chapter 4 Clinical Research 67
*Aziz Nather, Jamie Xiang Lee Kee &
Haitong Mao*

Chapter 5 Choice of Experimental Animals 83
*Aziz Nather, Jane Lim Jia Xin &
Elaine Yi Ling Tay*

Chapter 6	Cadaveric Research *Elaine Yi Ling Tay, Jane Jia Xi Lim &* *Aziz Nather*	115
Section III:	**Ethics and Statistics**	**147**
Chapter 7	Ethics for Research *Joy Le Yi Wong & Aziz Nather*	149
Chapter 8	Statistics for Clinical Research *Yiong Huak Chan*	181
Section IV:	**Writing Your Research**	**193**
Chapter 9	Tips for Scientific Writing *Claire Shu-Yi Chan, Wee Lin & Aziz Nather*	195
Chapter 10	Choosing the Right Journal *Wee Lin & Aziz Nather*	211
Chapter 11	How to Write an Original Research Article for a Journal *Wee Lin & Aziz Nather*	223
Chapter 12	Uncovering the Review Article *Zest Yi Yen Ang & Aziz Nather*	243
Chapter 13	Writing a Case Report *Zest Yi Yen Ang & Aziz Nather*	257
Chapter 14	Writing a Thesis or Dissertation *Zest Yi Yen Ang & Aziz Nather*	273
Chapter 15	What is Plagiarism *Eda Qiao Yan Lim & Aziz Nather*	283
Section V:	**Evaluating Your Research**	**295**
Chapter 16	Reviewing an Article *Aziz Nather*	297
Chapter 17	What Reviewers Look for in an Original Article *Joy Le Yi Wong, Wee Lin & Aziz Nather*	305

Chapter 18	Examining a Thesis or Dissertation *Aziz Nather*	309
Chapter 19	What Reviewers Look for in Writing a Thesis *Zest Yi Yen Ang, Joy Le Yi Wong & Aziz Nather*	319
Chapter 20	Objective Evaluation of Research Output *Aziz Nather*	325

Preface

This book is written as a step by step practical guide for all residents and young surgeons in the disciplines of Orthopaedics, Hand Surgery, Plastic Surgery, Vascular Surgery and all fields of General Surgery embarking on research, especially if they are doing so far the very first time. In addition, in Malaysia, Indonesia, the Philippines, Thailand and Singapore, a thesis is a compulsory requirement for a resident to complete "Masters in Orthopaedic Surgery". This book will also be especially useful to the Master's Degree student striving to complete his thesis.

In 1994, I was invited to conduct a two-day research seminar in University Science Malaysia by Dr. Hasim Mohamad, Head of the Department of General Surgery. The specific mission was to guide their Master's students to improve their theses. This seminar included closed door sessions — "research clinics" where each Master's students will present his thesis — whatever the stage they are in: proposal phase, on-going phase or completed thesis. It proved to be very useful. On the second day, two more "research clinics" were added to accommodate requests from Master's students in Obstetrics & Gynaecology and Orthopaedics. A similar two-day research seminar was conducted the following year. I noticed all students to be carrying a file of notes. On careful inspection, I found that these "notes" were actually hard copies of my overhead projection slides of the various lectures I gave the year before. The lectures included 'Planning Research', 'Choice of Experimental Animals', 'Writing up the

Finished Product', 'Writing up your Thesis', etc. I realised then that a book in this area is very much needed. It was only in 2000 that I found time to pen such a book titled 'Research Methodology in Orthopaedics and Reconstructive Surgery' published by World Scientific in 2002. This book took two years to complete.

This new book is an update of the 2002 publication. It is rewritten to focus on two important thrusts for research — a large section on planning and organising your research (Section II: Planning Your Research) and an equally large section on writing the research product (Section IV: Writing Your Research). The latter is often very little discussed. The young researcher is now more able to receive more instructions and guidance on how to plan his research. Still little advice is usually given on how to write up his research. This book is specially written to give practical guidance on how to write up their research projects.

The new book also includes an additional useful section (Section V: Evaluating Your Research) where the young researcher can also seek valuable insights into what reviewers look for in an article or into what examiners look for in examining a thesis.

I wish all residents and young surgeons every success as you embark on your research projects. Stay focused and committed and you will succeed.

Associate Professor Aziz Nather
Department of Orthopaedic Surgery
National University of Singapore
July 2014

About the Editor

- Associate Professor & Senior Consultant
  Division of Foot & Ankle
  University Orthopaedics
  Hand and Reconstructive Microsurgery Cluster
  National University Health System, Singapore
- Chairman, NUH Diabetic Foot Team since May 2003
- Medical Director, NUH Tissue Bank since Oct 1988
- Director, IAEA/NUS Regional Diploma Training Course for Tissue Bank Operators in the Asia Pacific Region
- Founding President and Current Honorary Secretary, Asia Pacific Association of Diabetic Limb Problems
- Chairman, ASEAN Plus Expert Group producing Guidelines on Management of Diabetic Foot Wounds Nov 2012– Nov 2014.
- Chairman, URGO Asia Pacific Wound Expert Board, April 2015– April 2017.

List of Contributors

Editor and Author
Aziz Nather
Associate Professor & Senior Consultant
Division of Foot and Ankle
Department of Orthopaedic Surgery
University Orthopaedics Hand and
　Reconstructive Microsurgery Cluster
National University Health System
NUHS Tower Block, Level 11
1E Kent Ridge Road
Singapore 119228

Chairman, NUH Diabetic Foot Team
Corresponding Member
International Working Group on the Diabetic Foot
The Netherlands
Medical Director, NUH Tissue Bank
Email: dosnathe@nus.edu.sg

9th NUH Diabetic Foot Research Team 2013
Haitong Mao
Jamie Xiang Lee Kee
Jane Jia Xin Lim
Elaine Yi Ling Tay
Department of Orthopaedic Surgery
NUHS Tower Block, Level 11
1E Kent Ridge Road
Singapore 119228

10th NUH Diabetic Foot Research Team 2014
Claire Shu-Yi Chan
Joy Le Yi Wong
Wee Lin
Zest Yi Yen Ang
Eda Qiao Yan Lim
Department of Orthopaedic Surgery
NUHS Tower Block, Level 11
1E Kent Ridge Road
Singapore 119228

Additional Contributor
Yiong Huak Chan
Statistician
Dean's Office
Yong Loo Lin School of Medicine
National University of Singapore
NUHS Tower Block, Level 11
1E Kent Ridge Road
Singapore 119228

Section I
Introduction

Lessons from Research: A Personal Experience

Aziz Nather

This overview is written from the Singapore Orthopaedic Association (SOA) Donald Gunn Lecture delivered on 10 October 2010 in Four Seasons Hotel during the 37th Annual Scientific Meeting of the SOA. The author was invited by then President of SOA, Dr. Inderjeet Singh. Professor P. B. Chacha (author's mentor) was invited to give a citation on the Donald Gunn Lecturer.

I was more than happy to accept Dr. Singh's kind invitation. Being a pioneer in the field of diabetic feet and bone banking both nationally and regionally, I naturally assumed that I was to be presenting on bone allograft or diabetic feet.

One can indeed imagine my surprise when Dr. Singh informed me that I was to be presenting on neither of the two topics! "Few people are interested in bone banking, and even fewer in diabetic feet!" Dr. Singh remarked. The question in my mind then was what else could I possibly be asked to present on?

The answer lay in research. As Dr. Singh rightly noted, everyone needs to do research. Being well known for my research, Dr. Singh hoped that I could give a good presentation on research so that medical officers and residents can learn from my experience. While it was a great privilege to deliver a lecture for the Donald Gunn lecture set,

I was initially also rather concerned. After all, the Donald Gunn lecture series is highly esteemed. There was a high repute to live up to, and I did wonder whether a lecture on research would have enough substance to merit such a big lecture.

Having never given such a lecture before, I spent a good six months of hard work preparing for the Donald Gunn lecture. In the end, I realised that the best way to approach such a topic would be from a personal point of view. With the Donald Gunn lecture having concluded four years ago, and many more similar lectures on research given by me the past few years, I have been inspired to write my own book on research to better inform residents and medical students on the often arduous, yet fulfilling process of research.

So, why research? Some say they want to leave a legacy behind for themselves through their research, while others argue that research is conducted for the betterment of mankind. For me, why I want to do research is motivated by a highly personal reason. It all really began on a passage to India ...

In 1980, I embarked on a research project to investigate the results of decompression for spinal metatheses with cord compression. It was found that in patients with cancers spreading to the vertebral bone pressing on the spinal cord, the outcome of spinal decompression varies. Moreover, the outcome of spinal decompression is dependent on several factors, including the degree of neurological deficit and type of cancer involved.

Having concluded writing the paper, my supervisor, Professor Kamal Bose, encouraged me to send it to Professor P. B. Chacha. I left the draft

A passage to India

Taj Hotel, Bombay

of the manuscript on Professor Chacha's table and by ten the next day, everyone in the clinic seemed to be talking about sending me to Bombay. I was astonished by the news: how could I possibly present a paper in Bombay when I have never even presented a paper before?

Anxious, I spoke to Professor Chacha about my concerns. Apparently, he was sending me to Bombay for an International Cancer Conference. The Chairman of the Singapore Cancer Society is Professor Chacha's friend and he was willing to sponsor me to Bombay for the International Cancer Conference at the Taj Hotel, Bombay.

Bombay

Bombay by night

I reacted to this sudden news with mixed emotions — elated to have been given such a valuable opportunity but apprehensive about whether I could deliver a lecture to a group of professors from all over the world. However, Professors Chacha and Bose reassured me, and ensured that they would support me all the way and render me with any help I could possibly require.

With their encouragements in mind, I headed off to Bombay with my wife and readied myself for the presentation I was about to deliver.

However, things took a turn for the worse at the conference when it was Professor Sen Gupta's turn to present his paper. Professor Sen's paper was critically challenged by students from another professor. I was very worried at that time — would I be able to back my paper up adequately? I remember telling my wife to pray for me; hopefully, the students would spare me!

Fortunately, when it was my turn, the students were all very kind and the presentation was a success! After the positive reception, my wife and I proceeded on to Jaipur, New Delhi and Agra for an extended holiday. That was when my wife had a sudden brainwave. She turned to me and commented on how we got to travel to so many places just from presenting a research paper. If I were to do more research and present more research papers, then we would get to travel more!

Jaipur   Red Fort, New Delhi   Taj Mahal, Agra

Indeed, my research work has brought me all over the globe – to exotic places ranging from Africa to South America, and Europe to Japan.

So, why do I do research?

The answer is simple: to travel and see the world!

Paladin once said, "Have gun, will travel". In my case, it is indeed "have research, will travel"!

Have Research,

Will Travel!

Of course, I have not told you that I love research. This activity excites me. I feel great as I add more clinical cases to my data collection for the research project; I perform the analysis with great interest and enthusiasm and I am thrilled with the results. Then comes the writing of the abstract for the research paper and the conference application. It feels superb when one's paper gets accepted. I am always thrilled to travel to yet another country.

I was born to do research!

Formula for Research (Fig. 0.1)

With years of experience, I would like to share with you my own formula for conducting research. My own successful approach lies with meticulous planning followed by brainstorming. Thereafter, I proceed to complete the research and present it in a good international conference to gain valuable feedback from the brainstorming with world-renowned experts. But it should not just end as a conference paper. One must persevere and write up the finished product. The

Fig. 0.1: Formula for research.

research process only ends when one finally achieves getting a publication in an international peer-reviewed journal with a good tier and high impact factor. Getting such a publication is the most difficult but most important step in the whole process.

I completed several research projects and succeeded in presenting several papers in many international conferences. Once presented, I also succeeded in writing most projects as articles in international and refereed journals.

Step 1: Planning the Research

Planning the "research process" is a detailed and meticulous exercise crucial to the success of the research plan (Fig. 0.2).

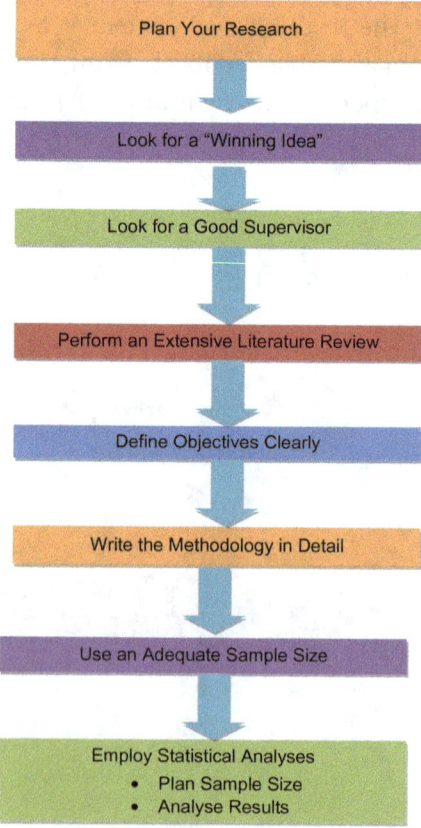

Fig. 0.2: Planning research.

It is important to first plan one's research properly, meticulously and in great detail.

Getting a "winning idea" — one that is of clinical importance — is critical. One needs a good clinical supervisor committed to research to share with one his research on a topic he has chosen.

It is then of paramount importance to do a literature review. After all, research cannot be done in a vacuum.

All these points will be discussed in greater detail in Section II of this book: "Planning Research".

Step 2: Brainstorming Research

It is always important to continually brainstorm the research one is conducting (Fig. 0.3) throughout the various different research phases: "Starting Phase", "Middle Phase" and "Completed Research". One can do this in a variety of different forums:

A good and very important forum is to present the research proposal to the department's quarterly research presentation.

It is a good avenue to get one's proposal to be critically analysed in detail by senior consultants and consultants experienced in research. Every part will be meticulously brainstormed — are the objectives clear? Are they feasible? Is the research with clinical impact? Is the methodology appropriate? Is statistical analysis adequately used? etc.

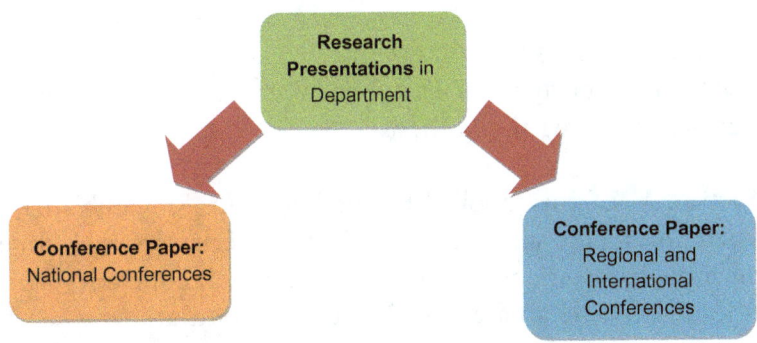

Fig. 0.3: Brainstorming research.

Such detailed brainstorming will give valuable feedback and shape the research process from the very start.

Next, it would be very useful to gain wider feedback from experts in the field of study by presenting the research paper in a national or international conference, preferably one dealing with that particular field of study. Of course at the same time, regional and international conferences provide one with travel grants and a good reason to travel to many more countries around the world!

Step 3: Writing the Final Product

One must not stop at presenting research at conferences. The research project would not be highly successful until one can obtain a publication in a Tier 1 or Tier 2 internationally refereed journal.

Section IV of this book: "Writing Your Research" is also devoted to writing one's research paper, with numerous tips on how to write papers for specific journals. These tips include:

Source: http://bibsandbaubles.com/wp-content/uploads/2011/09/writing-with-pen.jpg

- The style of English;
- Specific tips for original articles, review articles and case reports; and
- Specific tips for writing a thesis

Section V of this book: "Evaluating Your Research" has been designed to include chapters such as what reviewers look for in an article and what examiners look for in a thesis. It concludes with how one should evaluate an academic's "research output".

Some Final Advice

Jan–May 2014

10th Research Team (from left to right): Zest Ang Yi Yen, Chan Shu-Yi Claire, A/ Prof Aziz Nather, Lim Qiao Yan, Eda, Wong Le Yi Joy, Wee Lin

One must strategise and employ productive strategies for research. The most important advice I would leave for the reader is not to do research alone. Research is an arduous process that is at times challenging. It is productive to have a partner which will challenge your thinking every step of the way — someone to continue brainstorming your research process with you.

Better still, it is recommended that one joins "research engines" or form "research teams" to be more successful in research. By 'research engine', I mean that one should join focused research groups or areas of research such as a musculoskeletal tissue engineering group with a multi-disciplinary team including engineers, biological scientists, clinicians and research fellows.

If not, one could form a research team including research fellows, residents or research assistants. I myself have formed 10 research teams comprising research assistants over the past 10 years. I have resorted to employing research assistants (Year 0 Medical Students) with great success, since residents are in great demand and not readily available.

One must also seek help to source for the much-needed grants to do one's research. This is an art that must be learnt and mastered quickly. This issue will be discussed in greater detail in the chapter on procuring research grants.

I would like to leave this final piece of advice for all embarking on their research journey:

Research is lovely and exciting. Persevere. Stay Focused and Committed and you will Succeed. The joy you get when the research is successful and completed is well worth the trouble encountered!

I wish all residents and young surgeons every success as they embark on their research projects.

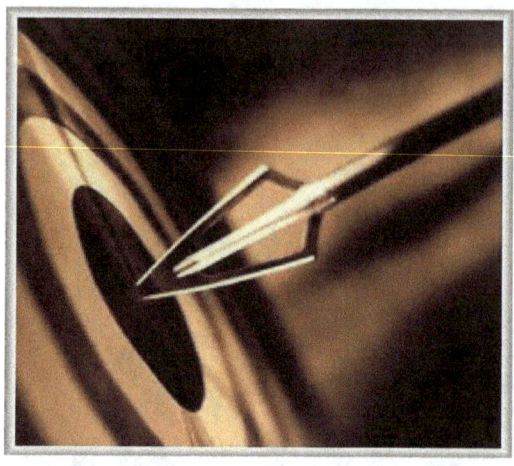

Source: http://www.macesports.com/arrow-target_flip.jpg

Associate Professor Aziz Nather

Editor

Section II
Planning Your Research

Chapter 1

Planning Research

Aziz Nather, Jamie Xiang Lee Kee & Haitong Mao

Choosing a Suitable Supervisor

"The first timer must not be thrown into the deep end of the swimming pool. He might just drown! ... He should have the guidance of a good supervisor."

— Professor P. Balasubramaniam,
Department of Orthopaedic Surgery, NUS

From the very start of any research project, a suitable supervisor should be present to provide guidance. A good mentor should be:

1. *An experienced clinician and keen researcher* — This places him in a good position to identify significant clinical issues meriting investigation. He can provide sound advice to the resident on the subject of his research project.
2. *Innovative and able to think outside of the box*
3. *Committed to mentoring the resident* — A young, inexperienced researcher should not be left to fend for himself.

Sourcing a "Winning Idea"

The most important part of planning is to decide on a "winning" project. The topic must be novel. The results of the research should have a significant clinical impact on medical practice. For this, we need an experienced clinician with a keen eye on choosing the right topic.

One example would be the concept of procuring stem cells from the filtrate bag of the Reamer Irrigator Aspirator (Fig. 1.1) by G. Cox, D. McGonagle, S. Boxall, C. Buckley, E. Jones and P. V. Giannoudis. This paper, titled 'The reamer-irrigator-aspirator (RIA): a harvester of mesenchymal stem cells (MSCs)',[1] was published in 2011 in the *Journal of Bone and Joint Surgery*.

Reamer-Irrigator-Aspirator

- Aspirated contents pass through the filter, trapping bony particles, and flowing into the "waste bag".
- The filtrate bag contains large numbers of MSCs. The effluent is potentially usable for the clinical transplantation of MSCs, without going through cell expansion in the laboratory. The expansion of cells will require two weeks in the laboratory.
- Studies show that Passage 2 cells from this effluent could differentiate into osteogenic, adipogenic and chrondrogenic lines.
- This is a revolutionary clinical find, which may well change the practice of tissue engineering.
- There is one other novel idea that has not been explored by the authors. The effluent is potentially a rich source of growth factors. Platelet-rich plasma (PRP) has been obtained by centrifugation of blood from the buffy coat layer.[2] By the same token, centrifugation of the effluent will give a good yield of PRP and other growth factors. This could also be explored and be an additional clinical application from the RIA.

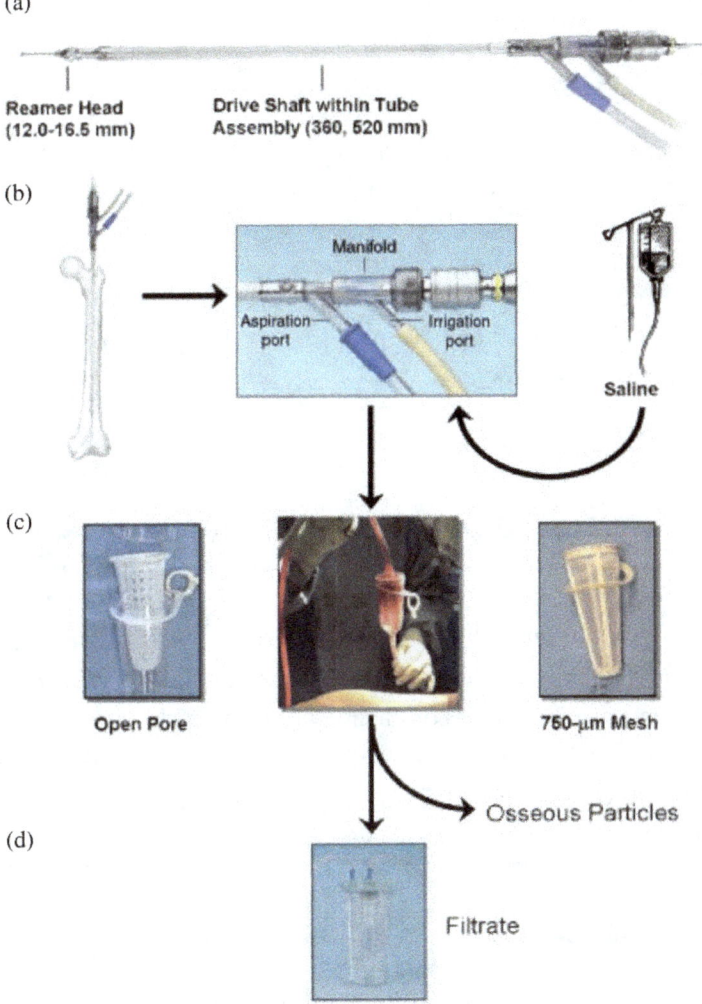

Fig. 1.1: A reamer-irrigator-aspirator.

In choosing the research topic, one must ask the following questions:

- *What is important about your project?* — Your research should be useful. It should aim to contribute to clinical practice.
- *What is new about your project?* — The topic must be original. If there has been previous work done on the topic, your research

should seek to offer an angle that has not yet been addressed by others.
- *Is the idea viable?* — The research should be able to be conducted in the facilities available to the researcher.
- *Will the costs fit within the budget given?* — The researcher should estimate the budget costs to make sure that it is within the budget allocated for the research.
- *Can the project be completed in time?* — Time is a very important factor. The planning must take into account the time needed to complete the project. Many projects are left uncompleted due to insufficient time allocated.

PLANNING AHEAD: DO YOU HAVE ENOUGH TIME?

In-vivo animal experiments studying the biology of healing of tendons, cartilage, and bone require at least 2 to 3 years for completion.

In-vitro biomechanical studies on cadavers or on animals require less time. It is possible to complete within a 1- to 1.5-year frame.

With clinical studies, the time required is generally less than animal experiments.

Writing Your Objectives

After selecting the research topic, one should first write clearly the detailed objectives of the study.

Formulating Your Hypothesis

One should formulate the hypothesis being studied.

A hypothesis is a speculation which will be either proved or disproved according to the evidence. It has to be testable, and has to be formulated early on in the planning stage.

Reviewing Literature

Research must never be done in a vacuum. One should not assume a specific clinical problem to have never been researched before. A thorough review of all literature on the problem should be performed to ensure the topic is a novel one.

How to Review Literature

Fig. 1.2:   General process of a literature review.

Reviewing literature is an art to be mastered. It must be extensive. This process involves the following stages (Fig. 1.2):

- Procurement of articles
- Reading and critical analysis of articles
- Ranking of articles based on relevance
- Summarising the salient points of each article
- Filing of articles for future reference

Appraising Articles with 10 Questions

The following 10 questions should be asked when critically analysing a research article.[3]

1. *Is the study question relevant?*
   The study should address the topic that you are researching on and add to what is already known about that subject.

2. *Does the study add anything new?*
   While most papers may not make a substantive new contribution to existing knowledge, research papers that make an incremental advance can also be of value. For instance, an article may increase

the validity of previous research by replicating its findings. It may also extend original findings to new populations of patients or clinical context.

3. *What type of research question is being asked?*
   Identifying the specific research question addressed by the article is the most fundamental step in article analysis. There are generally two kinds of clinical research questions: questions about the effectiveness of treatment and questions about the frequency of events such as incidence of diseases.

4. *Was the study design appropriate for the research question?*
   Meta-analyses of randomised controlled trials (RCTs) and individual RCTs are most suitable for studies that answer questions about effectiveness. Meanwhile, observational studies are the most appropriate for questions about the frequency of events.

5. *Did the study methods address the most important potential sources of bias?*
   The bias may be a random error attributed to chance. Alternatively, it may be a systematic bias that is inherent in the study methods.

6. *Was the study performed according to the original protocol?*
   If a study deviates from the planned protocol, its validity or relevance can be affected. A failure to recruit the planned number of participants, for instance, reduces the power of the study to demonstrate significant findings. Other possible deviations include changes to inclusion and exclusion criteria, employed techniques and duration of follow-up.

7. *Does the study test a stated hypothesis?*
   The study hypothesis should be identified before the study is conducted. The study may otherwise be prone to false-positive findings. In addition, check that all data relevant to the stated study objectives have been reported, and that selected outcomes have not been omitted.

8. *Were the statistical analysis performed correctly?*
   All quantitative research articles should explain the tools used in the statistical analysis and the rationale for them. The approach to dealing with missing data and the statistical techniques applied should also be specified. Additionally, patients lost in follow-up and missing data should be clearly identified.

9. *Do the data justify the conclusions?*
   The conclusions should be reasonable based on the data collected. Make sure that no overemphasis is placed on statistically significant findings that in reality are differences too small to be of clinical value.

10. *Are there any conflicts of interest?*
    Examples on such conflicts include receipt of salary or consultation fees from the company that has sponsored the research, patents related to the research and industry funding for educational events, travels or gifts. Most journals require authors to declare any potential conflicts of interest. It is up to the reader to decide if the declared factors are significant enough to influence the validity of the study's findings.

Ranking Articles According to Type of Research

Each article should be ranked according to several factors, namely: the type of article, type of journal, year of publication and institution of authors.

Type of article

Systematic reviews, meta-analyses and RCTs tend to be the most objective studies with the least potential for bias. They also provide the strongest evidence (Fig. 1.3 and Table 1.1). Multi-centre studies provide more significant results than single-centre studies.

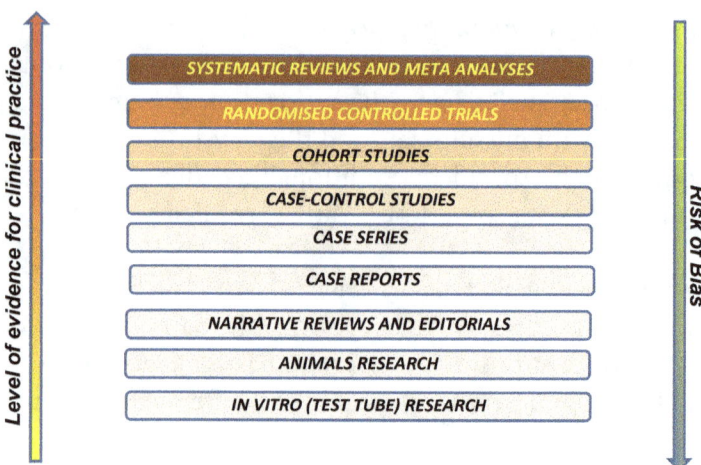

Fig. 1.3: Ranking of studies.[4]

Table 1.1: Hierarchies of evidence for questions of therapy, prevention and aetiology.[5]

Level	Type of Study
Level 1a	Systematic review (with homogeneity) of RCTs
Level 1b	Individual RCT (with narrow coincidence interval)
Level 1c	All-or-none studies
Level 2a	Systematic review (with homogeneity) of cohort studies
Level 2b	Individual cohort study (including low quality RTC, e.g. <80% follow-up)
Level 2c	"Outcomes" research; ecological studies
Level 3a	Systematic reviews (with homogeneity) of case-control studies
Level 3b	Individual case-control study
Level 4	Case series (and poor quality cohort and case-control studies)
Level 5	Expert opinion without explicit critical appraisal, or based on physiology, bench research or "first principles"

Type of journal

Articles published in high impact, internationally refereed journals are more significant than those published in regionally or locally refereed journals (Table 1.2). It is also better if the journal is specialised in the particular field of medicine you are researching.

Table 1.2: Journal rankings.

Ranking	Type of Journal
1	Tier 1 Internationally Refereed Journal
2	Tier 2 Internationally Refereed Journal
3	Regionally Refereed Journal
4	Locally Refereed Journal

Year of publication

Articles published recently should be procured.

Institution of authors

The international standing of the author's institution should also be taken into account.

Preliminary write-up of discussion and introduction

Based on the literature review, one first begins to design a discussion on the topic. This will include evidence for and against the hypothesis. The references in support or against the hypothesis must be appropriately quoted. This is best done immediately after the completion of the literature review, when both the ideas and facts are still fresh in the mind. The literature review and subject discussion are the most tedious to finish in the writing of the article. Possessing a completed discussion before one embarks on the research project will be very useful in writing the final manuscript several months later.

Once the discussion has been written, one can then select the information that can best be used for the introduction. The introduction summarises what is known on the topic and what is not. It ends with the objectives of the study.

This will be discussed in detail in Chapter 11.

A useful procedure to follow

1. Write objectives clearly.
2. Formulate hypothesis.
3. Review literature.
4. Plan discussion.
5. Plan introduction.

Considering Ethics

One must always bear in mind ethical considerations when designing a research project. This is true for all types of research: clinical, cadaveric or otherwise. All research projects submitted to the National Medical Research Council (NMRC) in Singapore must first be evaluated and approved by the Ethics Committee before it can be approved for funding by the Council. Ethics will be further elaborated upon in a later chapter.

Statistical Planning

It is essential to involve a statistician early in one's project. This is especially true when one is designing a prospective clinical study.

What a statistician can do:

- Advise on the size of the study population to be adopted for the project to be clinically significant
- Decide how to select a representative sample after enumerating all variables that could influence the results
- Evaluate and determine the statistical significance of the results, upon completion of the study (Fig. 1.4)

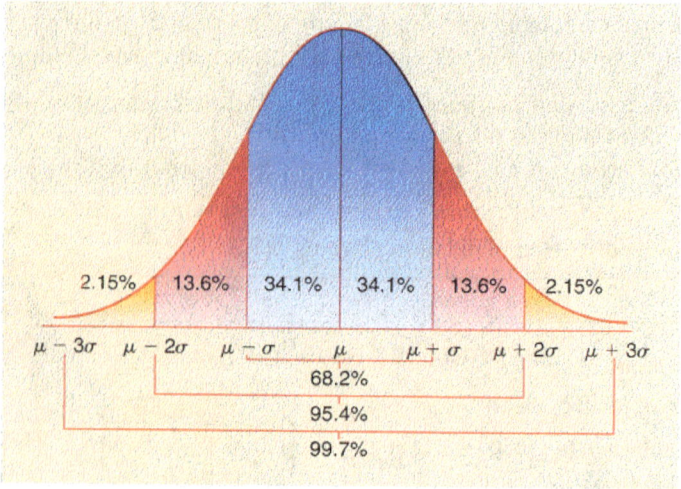

Fig. 1.4: A normal curve can be used to determine the statistical significance of experimental results.

Brainstorming Research

Upon completion of the initial research proposal, it should be presented at departmental level for detailed discussion and constructive criticism. Based on the feedback obtained, the research proposal should be rewritten and presented to the department again. Repeated discussions allow problems unanticipated by both the researcher and the supervisor to be raised. Be it work at the proposal stage, on-  going work, or completed research, residents should presented their research to the department for peer review every three months.

If possible, research should be presented at International Research Meetings so as to obtain feedback from a wider, international pool.

Pilot Study

A pilot study is an essential small scale preliminary study conducted in order to evaluate feasibility, time, cost, adverse events and effect size (statistical variability). It is a method to predict an appropriate sample size and improve upon the study design prior to performance of a full-scale research project.

Using a pilot study, one can find out all the practical problems that will arise. Such problems include:

For clinical reviews:

- Mechanism of recall for patients
- Clinic space to conduct the review

For cadaveric research:

- Shortage of proper surgical instruments
- Lack of storage facilities for proper dissection of cadavers in cadaveric research

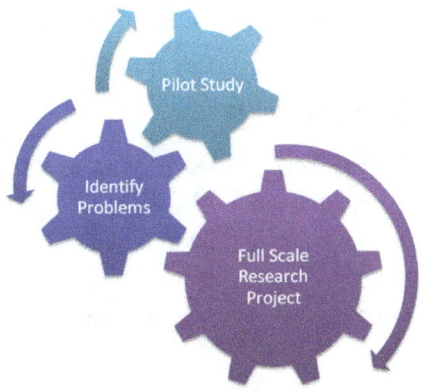

For animal experimentation:

- Anaesthesia for the animals
- Operative techniques
- Post-operative care of the animals
- Lack of proper surgical instruments

The research protocol can then be written to fully address the problems surrounding the project, based on the pilot. The pilot study is especially important for animal experimentation.

The pilot study has to show that the research project could run smoothly without any major problems. Only then is the researcher ready to embark on the final writing up of the research proposal for a full-scale research project application.

Research Groups

A research group consists of 3 to 4 principal investigators (PIs) who share the same research interest, e.g., tissue engineering. One PI may be interested in the effect of MSCs on bone, the other on cartilage and the third on tendons. It is strategic for them to team up to brainstorm regularly. They may have bioengineers dealing with scaffolds and tissue engineering specialists dealing with cell culture and growth factors. By forming a research group, experience gained in one tissue may be beneficial to the PI dealing with another tissue. Each PI should ideally possess a research grant and have a research fellow employed to work for the project.

Research Fellow

A research fellow is very useful for conducting a research project, especially when the PI is a clinician heavily involved in patient work. The research fellow is available to do the research project on a full-time basis. He can run the clinical trials on patients or perform the animal experiments for basic research. The fellow is responsible for planning and coordinating the research project. He performs all the documentation for the research and analyses all the results obtained. He is also responsible for writing up the article to be produced from the project, including literature search, discussion, etc. He will produce progress reports and final reports, as well as run all administration required. His work should also involve writing up future research grant applications.

To engage a research fellow, the PI should have a grant of $250,000 or more. A research fellow (post-doctorate Ph.D.) will require at least $60,000 per annum.

Do not wait until a big grant from, e.g., NMRC, A*STAR's Biomedical Research Council (BMRC), is obtained before hiring a fellow. One must think out of the box. Two to three clinicians can group up to hire a Ph.D. fellow for their research by contributing $2,000 monthly each. The research fellow can write up the application for research grant and run the research project efficiently for the group. This is a good investment. The group can produce good research work with a full-time research fellow.

Residents

The group can also include residents keen to pursue research. This is a win-win situation as both the consultants and residents achieve great satisfaction from the publications arising from the research. The resident will benefit from research to advance his career whilst the consultant may get the promotion he desires.

Students

Junior College (JC) students who have completed A levels (in November), while waiting for medical university enrolment, can also be recruited into the research group. The senior author employs four JC students full time for five months each year from January to May as research assistants.

8th Research Team 2012

9th Research Team 2013

10th Research Team 2014

References

1. Cox, G., McGonagle, D., Boxall, S., Buckley, C. T., Jones, E. & Giannoudis, P. V. (2011). The reamer-irrigator-aspirator (RIA): a harvester of mesenchymal stem cells (MSCs). *J. Bone Joint Surg. Br.* **93**(4): 517–524.
2. Dasde, S., Manohara, R. & Nather, A. (2005). Platelet-rich plasma in orthopaedic surgery: basic science and clinical applications. In: Nather, A. (ed.) *Bone Grafts and Bone Substitutes.* New Jersey, London, Singapore: World Scientific Publishing, pp. 387–403.
3. Young, J. M. & Solomon, M. J. (2009). How to critically appraise an article. *Nat. Clin. Pract. Gastroenterol. Hepatol.* **6**(2): 82–91.
4. Bhandari, M. & Joensson, A. (2008). *Clinical Research for Surgeons.* New York: Thieme Medical Publishers, Inc., p. 39.
5. Hemingway, P. & Brereton, N. (2009). What is a systemic review? Hayward Medical Communications, Retrieved from http://www.medicine.ox.ac.uk/bandolier/painres/download/whatis/syst-review.pdf

Chapter 2

Procuring Research Grants

Haitong Mao & Aziz Nather

Types of Research Grants

There are many types of grants, both local and international (Fig. 2.1), available to the researcher. One should choose to apply to the grants most appropriate to their research area.

Local Grants

A*STAR

Aside from scientists in its research institutes, the Agency for Science, Technology and Research (A*STAR) also provides grant support and research resources to other publicly funded institutions such as universities, hospitals and specialty centres in Singapore.

Source: http://www.a-star.edu.sg/

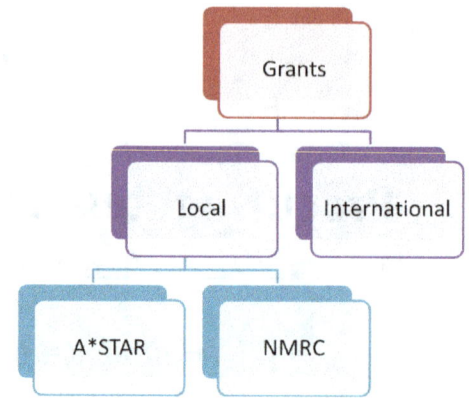

Fig. 2.1: Overview of the types of grants available.

As of 2013, its various calls for grants and sponsorships that biomedical researchers can apply include:

Bilateral Joint Research Grant Calls (International)

The grants under this category are jointly awarded by A*STAR and another country. The project needs to be two principal investigators (PIs). One investigator should be based in Singapore, and the second based in the country awarding the grant.

BMRC General Grant Calls

The Biomedical Research Council (BMRC) funds basic biomedical and translational clinical research that is relevant to human health as well as research that looks at the causes, consequences, diagnosis and treatment of human diseases. Each year, the BMRC and the Ministry of Health (MOH)'s National Medical Research Council (NMRC) conduct a joint grant call for investigator-driven biomedical research proposals. The grant call typically opens on the first working day each May and remains open for no less than one month.

BMRC Consortia Grant Calls

The BMRC Consortia Grant Calls encourages collaborations between A*STAR research institutes and consortia and the extramural community in strategic thematic areas through their consortia grant calls. Consortia in the BMRC (namely the Singapore Bioimaging Consortium, Singapore Immunology Network and Singapore Institute of Clinical Sciences) conduct calls for proposals for funding annually.

SERC Public Sector Funding

The Science and Engineering Research Council (SERC) 2012 Public Sector Research Funding (PSF) Grant Call seeks proposals (1) in areas of upstream research and/or (2) exploring novel concepts (rather than application of existing approaches to new technologies or development of systems). Projects in the field of bioengineering are preferred.

SERC Biomedical Engineering Programme (BEP)

The BEP seeks to foster clinician–engineer collaborations to develop medical devices and solutions to clinical problems. In particular, it supports collaborative research projects with emphasis on devices, procedures, diagnosis and clinical systems to improve patient care and cost efficiency of the healthcare system.

A*STAR-CIMIT Joint Call for Proposals

Since January 2010, A*STAR and the Center for Integration of Medicine and Innovative Technology (CIMIT) have collaborated on growing Medical Technology (MedTech) innovations and activities between Singapore and Boston. For this grant, Singapore–Boston collaborations are encouraged.

The types of grant available and the specific requirements of each grant call vary from year to year. Details can be found online, including on A*STAR's website.

NMRC

The NMRC provides research funds to healthcare institutions and awards competitive research funds for individual projects.

Source: http://www.ntu.edu.sg/home/quanliu/research.htm

CS Individual Research Grants (CS-IRGs)

The goal of this grant is to enable individual clinician-scientists (CS) to carry out medical research on a specifically defined topic for a period of three years in local public institutions.

CS-IRG New Investigators Grant (CS-IRG-NIG)

The CS-IRG scheme has a sub-category for new clinical investigators to attract and nurture young local talents to pursue competitive and innovative translational and clinical research. Applicants with substantial research experience *are not* accepted under this category. The applicant cannot apply to other funding sources as a PI or co-PI for the same evaluation period. He also must not have held any national grants (e.g., from NMRC, A*STAR, NRF, MOE AcRF Tier II, etc.) or international grants (e.g., from MRC, NIH, NHMRC, etc.) as a PI or co-PI.

Source: http://www.finaid.ucr.edu/typesAid/Pages/grants.aspx

Cooperative Basic Research Grants (CBRGs)

This grant is provided to **non-clinical researchers** to conduct research proposals in basic and translational clinical research that are relevant to human health as well as research that looks at the causes, consequences, diagnosis and treatment of human diseases. The proposed research project must be based in Singapore.

CBRG New Investigators Grant (CBRG-NIG)

The CBRG scheme also has a sub-category for new non-clinical investigators who have not held a national or international grant. As with **CS-IRG-NIG,** applicants with substantial research experience *are not* accepted under this category.

Bedside & Bench (B&B) Grant

This grant aims to foster closer interactions between the basic scientist and the clinician to translate scientific discoveries in the laboratory to clinical useful and commercially viable applications to improve health outcomes. The B&B Grant will fund collaborations between two co-PIs, one of whom is a basic scientist and the other a clinical investigator. Projects such as experimental medicine studies/phase zero studies or projects that will lead to the development of novel biomarkers, medical devices, therapeutics, new product candidates, technologies or techniques that can improve healthcare outcomes will be considered for funding.

Clinical Trial Grant (CTG)

The CTG is intended to support clinicians in carrying out clinical trial studies for the development of novel therapies for healthcare needs. There are three schemes under the CTG programme, namely the (1) Co-Development Scheme which supports clinicians who wish to collaborate with the industry, and (2) Early-Phase and (3) Late-Phase Schemes of the Investigator-Initiated Trials which support clinicians who wish to conduct clinical trial studies on therapies of their own interest.

Translational & clinical research (TCR) Flagship Programme Grant

The TCR Flagship Programme Grant aims to bring together the best complementary research strengths in hospitals and national disease centres, universities and A*STAR research institutes to focus on disease or research themes of strategic importance. Collaborations across institutions are encouraged. The grant is funded over five years.

Health Services Research Competitive Research Grant (HSR CRG)

The HSR CRG is a MOH research grant established in 2009. This CRG aims to promote the conduct of HSR and enable the translation of HSR findings into policy and practice. HSR CRG is provided to individual researchers (the PIs) to enable them to carry out HSR on a specifically defined topic within a defined time period (two years). The focus of the research should be translational in nature.

Health Services Research New Investigator Grant (HSR NIG)

The HSR NIG is a sub-category of the HSR CRG, which aims to help new HSR researchers take initial steps towards obtaining their first independent national-level grants. Experienced researchers should not apply for this grant.

International Grants

Researchers can also consider applying to international organisations for grants.

AO Foundation

The AO (*Arbeitsgemeinschaft für Osteosynthesefragen*, or conducting research in bone healing) Foundation provides start-up grants that support basic scientific, pre-clinical and clinical research in all areas of trauma, surgery of the musculoskeletal system and related problems. It provides predominantly seed money to individual researchers and research groups, finances pilot studies and supports investigation of new and unconventional ideas or hypotheses. This grant category is designed to encourage young investigators as well as experienced researchers submitting novel high-risk projects.

Sponsorship

Commercial companies

The researcher can also collaborate with commercial companies that may be willing to provide sponsorship for projects that involve their products.

Writing Up a Research Grant

How a research application is written is an important factor deciding whether the proposal is considered for funding or not. Usually an organisation will receive many research applications. All applications received for each period of review will be carefully evaluated by members of a research  committee. A poorly written application reflects insufficient preparation on the part of the PI and his team.

Considerable planning and preparation must be done before one is ready to write a good research application. The format to be adopted varies depending on the application form provided by the organisation from which one is applying the grant. The basic considerations for all applications, however, remain the same.

One should address the following areas in the application:

- Names of applicants
- Title of research proposal
- Duration of research proposal
- Abstract of research proposal
- Specific aims

- Significance of project
- Preliminary studies
- Methodology
- Budget estimates and justification
- Other support
- Biographical data of applicants
- Suggested names of reviewers

Applicants of Research Proposal

It is crucial to identify the PI when planning the research project. The PI plays a key role in all aspects of the project, and is responsible for the actual writing up of the research application for submission.

The PI must:

- Have the capability and experience to carry out the research to its completion
- Be committed to perform the research for the whole duration of the project, usually a minimum of 2 to 3 years
- Be available and not on sabbatical leave or overseas attachment during the whole duration of the research project. Before the grant is disbursed, the PI is usually required to sign an undertaking that for the period of the project, he has no plans to leave the university or the hospital.

The PI also chooses other members of his team for the research project. Each member must have a specific role to play in the project and must be able to assist the PI. A multidisciplinary team is preferred, since preference is usually given to multidisciplinary research projects in the allocation of funds for research grants.

Title of Research Project

The title of the research project makes a significant impression on the members of the review panel. The title should be concise and "catchy" to gain the attention of the reviewers. The significance of the project should be reflected in the title. It is best to carefully decide the title after the specific objectives, detailed methodology, etc. are written.

Abstract of Research Project

The abstract of a research application is perhaps its most important part. It must serve as a concise and accurate description of the research proposal (Fig. 2.2), and should be written last.

The abstract is often the first thing that reviewers will read. Thus, it creates an important first impression on the reviewers' minds and they proceed to read the rest of the research proposal.

Fig. 2.2: Diagrammatic representation of what an abstract should include.

Specific Aim

It must be very clear to the reviewer what the specific objectives the research intends to accomplish. The hypothesis tested should also be clear, to show that the research is focused towards studying specific objectives.

Significance

The background of the research proposal serves as an introduction to the clinical problem being addressed in the proposal. A detailed review of literature addressing the clinical problem, properly referenced, should be written. The discussion must summarise the work already done on the topic, the gaps existing in the current knowledge and how the current research study could fill such gaps (Fig. 2.3). The clinical significance of the research should be elaborated. Relevant references should be appended at the end of this section.

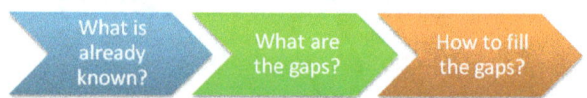

Fig. 2.3: Topics to be included under discussion.

Preliminary Studies

It is very important to provide an account of the PI's preliminary studies (if any) pertinent to this current research proposal. Such studies will help to establish the experience and competence of the investigator to conduct the research. If the PI has no experience at all in the field of research, his credibility would be seriously questioned. In such cases, it would help to mention if one of the collaborators has considerable experience in the field.

Methodology

The experimental design and the procedure used to accomplish the specific objectives of the project must be written and discussed in detail. One should include:

1. The protocols to be used;
2. The plan of the experiment;

Example

In a research proposal "Role of BMP in enhancing anterior lumbar interbody fusion using allografts in non-human primates", it will enhance the chances of the application if the PI writes:

1. He has done previous work on bone allografts and is either running a bone bank or has access to it.
2. He has been doing spinal surgery for at least 2 to 5 years and is therefore well-qualified to perform the spinal surgery required for the project.
3. He has previously worked on monkeys. If not, he should have at least worked on animal experiments before using such animals like rabbits, cats, dogs, pigs or sheep and has access to facilities for anaesthetising and looking after monkeys post-operatively. There should also be a veterinarian available to look after the anaesthesia and post-operative care of the monkeys.

3. The parameters to be studied;
4. The methods or means by which the results or data will be analysed;
5. All potential difficulties and limitations of the experimental design, procedure, protocol, parameters studied and data evaluation; and
6. All hazards to personnel in carrying out the research and precautions taken to overcome such dangers.

Animal experimentation

For basic research involving *in-vivo* animal experimentation the protocol must include the following:

- Experimental model/design chosen
  The investigator must describe in detail the experimental design chosen and the particular animal selected for the research (Fig. 2.4). Illustrations carefully drawn and labelled are extremely useful to demonstrate clearly the experimental design adopted.

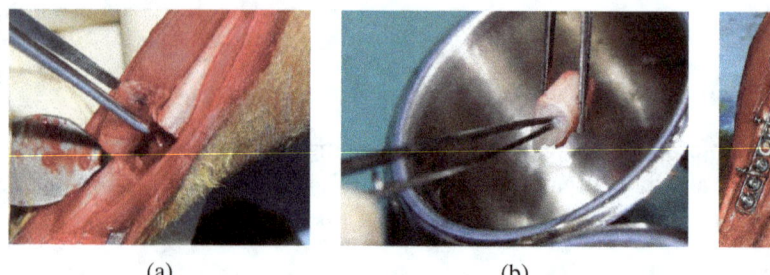

Fig.2.4: An experimental design of allograft-MSC transplantation into a tibial defect of a rabbit. a: A 1.5-cm cortical segment of the mid-tibia in a rabbit being excised; b: cortical allograft being filled with autologous MSCs from the rabbit; and c: the allograft with stem cells replacing the defect and reconstruction done with a plate.

- Anaesthesia
  The type of anaesthesia chosen for the animal must be described, e.g., intravenous ketamine administered in the ear lobe marginal veins of rabbits, or general anaesthesia with intubation for large animals such as dogs, pigs, sheep and monkeys. The availability of a veterinarian is useful especially for performing general anaesthesia for the larger animals.
- Operative procedure
  The surgical technique employed must be detailed, starting with the administration of prophylactic antibiotics parenterally and shaving and cleansing of the animal using a motorized shaver. After cleansing with povidone/iodine and applying sterile drapes, the incision made is described and the operative procedure performed outlined in detail (Fig. 2.5). It includes closure of the skin. It ends with the post-operative care of the animal for pain and prophylaxis against infection.
- Post-operative care
  One should record the analgesics delivered post-operatively mixed with the chow for the animal to ensure good pain relief. One should also include the oral antibiotics administered to reduce the infection rate.

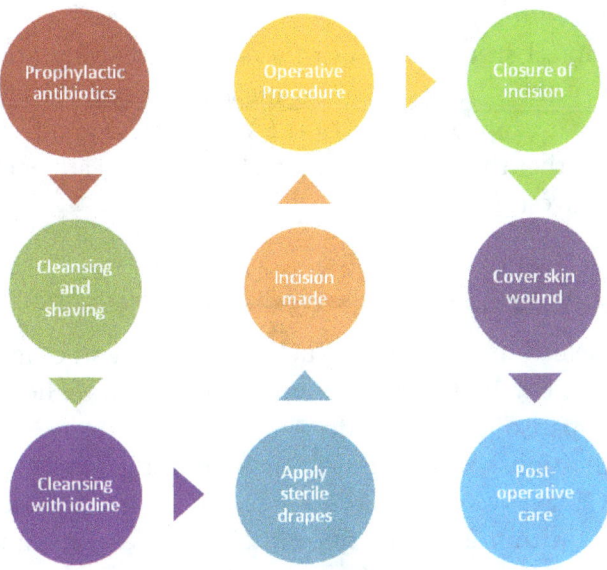

Fig. 2.5: Flow chart for an operative procedure.

- Controls
  The controls for the experimental design must also be carefully designed. The validity of a research project critically hinges on the selection of proper controls.
- Plan of experiment
  The plan for the whole experiment including experimental model and control must be outlined. In animal experiments, at least six animals are usually required for each observation period.

  The total number of animals needed from the experiment is calculated allowing for about 20% wastage due to infection or mortality.
- Parameters studied
  The parameters to be studied must also be described in detail. For example, for studies on fracture healing the parameters include plain radiographs (anteroposterior and lateral views), macroscopic examination of bivalved specimens and histological examination of specimens using undercalcified or decalcified sections of 10 μm thickness stained with haematoxylin and eosin or with other stains.

- Other procedures

    Biomechanical studies can also be performed using animals. The selection of the study population must be stringent to ensure that all animals are approximately of the same age, sex and weight. It is mandatory that the experiments on all specimens are performed by the same investigator.

Cadaveric research

Criteria for selection of cadavers (such as age, sex, and ethnicity) must be recorded in detail. The study population should be homogeneous for valid comparisons to be made.

The conditions of storage of the cadaver parts must also be described. Cadaveric parts used for research must be those procured from patients within 24 hours of death. The specimens must be kept in a freezer at –40°C (Fig. 2.6). Specimens used for testing must be cadaveric parts stored within two weeks of procurement.

Cadaveric studies take approximately 1 to 1.5 years. One should take into account the time needed to procure cadaveric parts.

The Instrom machine, or other machines used, must be described in detail. The type of biomechanical testing performed must also be accurately outlined, including type of jigs and type of test performed. The parameters of biomechanical testing recorded must include strength, stiffness and energy of absorption (Figs. 2.7 and 2.8).

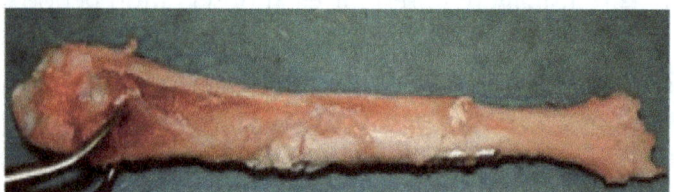

Fig. 2.6: Frozen tibia from a cadaver stored in a freezer at –40°C before use for cadaveric research.

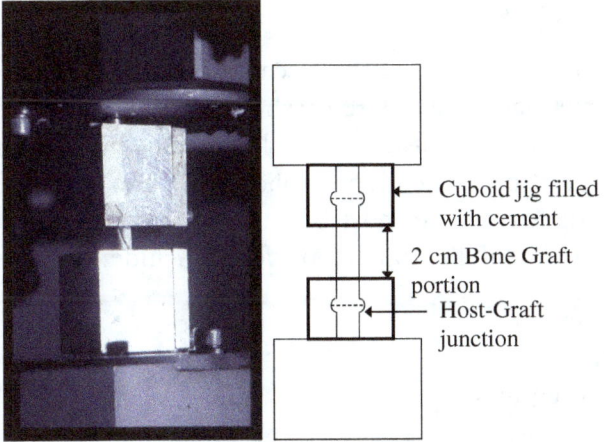

Fig. 2.7: Torsional testing of tibia allografts in an adult cat.

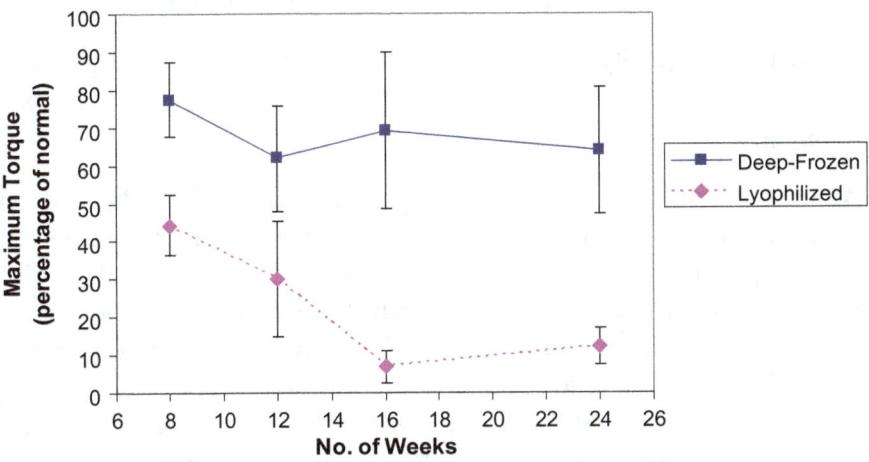

Fig. 2.8: Maximum torque of deep-frozen *vs.* lyophilised grafts in an adult cat.

Clinical studies

Prospective studies are preferable to retrospective studies. In designing a clinical study, a statistician should be involved from the start. This ensures that the cohort size is large enough to be statistically significant. Proper controls must be included in the research protocol.

Budget Estimates

This is an important part of the research proposal. It is useful to seek help from the department's research administrator in writing up the application. The administrator could also help to coordinate the research project once the grant application is successful.

The budget estimate must be realistic and take into account all known costs. The items to be considered include:

- Manpower
- Equipment
- Materials required
- Operating costs
- Cost of drugs/antibiotics/analgesics/dressings
- Other consumables
- Miscellaneous items, including photography and transport costs

One must justify large costs such as manpower and equipment.

Manpower

Manpower often constitutes the largest item in funding. It is crucial to apply for the necessary manpower to make sure that the research can be performed smoothly. Doctors are usually busy with clinical services, leaving little time for research. One should therefore recruit a full-time hire employed on a contract basis for performing the research: a research fellow, research assistant or technologist.

It is best to request for a post-doctoral fellow or at least a research assistant or technologist who will be able to perform the research work.

Equipment

It is important to budget for the necessary equipment and surgical instrumentation required.

For example, in the experiment on lumbar interbody fusion in monkeys using BMP-2 (bone morphogenetic protein 2), it is important

Who should you employ?

 With a large grant ($250,000 or more)

   Research Fellow

    Post-doctorate (Ph.D.) at $60,000 per annum

 With a small grant (less than $250,000)

   Technologist (Diploma) at $35,000 per annum

   Research Assistant (Degree) at $40,000 per annum

   Staff Nurse at $30,000 per annum

   Assistant Nurse at $15,000 per annum

to budget for a basic orthopaedic set, a spinal instrument set and an osteotome instrument set. The individual items in the sets required could be obtained in consultation with the operating theatre nurse. As this is major surgery, a diathermy machine must also be indented for the experimental surgery.

For biomechanical studies, a Shimadzu Universal Testing Machine (Fig. 2.9) or an Instron Machine must be available. The cost of this item is about $250,000! It is useful to collaborate with a bioengineer using such equipment. Additional costs include cost of equipment to mount specimens, jigs and cement. In other biomechanical experiments, transducers must be budgeted for.

For cadaveric research, it is essential to budget for at least one −40°C freezer for storage of the cadaveric parts. Instruments for cadaveric dissection including special retractors, portable oscillating saws, etc. must be indented. Costs of jigs for special mounts must also be included.

Fig. 2.9: (from left to right) **Shimadzu Universal Testing Machine; Instron Machine; −86°c freezer.**

Adapted from: http://www.ssi.shimadzu.com/products/product.cfm?product=ag-x
http://www.melbtest.com.au/fatigue.php
http://www.torontech.com/materials-testing/industrial-testing/ultra-low-temperature-laboratory-freezer

Materials and supplies

The cost of experimental animal to be use must be estimated to include the cost of maintaining the animals throughout the observation period. The maintenance cost far exceeds the cost of the animal to be procured. For example, the cost of a New Zealand white rabbit is about $70, but cost of maintenance per month is about $50.

Consumables

Items to be budgeted for include:

- Anaesthetic items
  - Anaesthetic drugs, syringes, needles, normal saline, gauze, etc.
- Surgical items
  - Gloves, masks, povidone/iodine, prophylactic antibiotics, normal saline, gauze, blades, sutures, etc.
- Histological items
  - Fixatives, decalcifying agent, dehydrating and clearing agent, embedding substances, glass slides, cover slips, stains, etc.

- X-rays
    - Films, processing chemicals
- Photography
    - Slides, prints

> ### *How long should you plan for?*
>
> For better planning of the research project and to facilitate disbursement of grants when it is approved, the researcher must also estimate actual costing to be requested for each year period.
>
> For manpower costs, the projection must include annual increments for Year 2 and for Year 3 for a three-year project.
>
> For equipment, most equipment needs to be purchased in Year 1.
>
> For Materials and for Miscellaneous Costs, a good principle is to ask for one-third of the total funds yearly for a three-year project. For a two-year project, ask for half the total funds for each year.

Budget estimations for clinical research

In planning clinical research protocols, budgetary considerations include:

- Manpower requirements
  It is important to request for a clerk or research assistant to co-ordinate the clinical research protocol. An experienced staff nurse or assistant nurse will be appropriate for the task. This person will be responsible for:
    - Tracing medical records and X-rays
    - Arranging appointments
    - Completing documentation, including assisting patients to fill in assessment protocols for the project
    - Obtaining consent from the patient to participate in the research
- Costs of outpatient fees
  For research, outpatient fees for patients should be waived.

- Costs of additional investigations required
    - X-rays
    - CT Scans
    - MRI
    - Blood tests
    - Culture and sensitivity
    - Histology

    The costs of these investigations must also be provided by the research grant.
- Costs of drugs, surgical implants or other treatments for clinical trials

    For clinical trials investigating the outcome of treatment, e.g., a new drug or surgical implant, the costs of the treatments used must also be provided. One should source for sponsors — commercial companies dealing with the drug or surgical implant. Once sponsorship is secured, the research application is more likely to be approved.

Case studies of budget estimates

Two examples of actual costing of budget estimates for two different projects are presented to demonstrate the principles involved in writing a budget proposal.

The first is the actual costing for Project 1, titled 'Biology of healing of osteochondral composite allografts in rabbits'. The project life is three years, as shown in Table 2.1.

In this application, the manpower costs amount to about 38% of total project costs (Fig. 2.10). This is because only a technologist is requested for. While the application will be more likely to succeed, it must be remembered that the PI does not have the support of a more experienced research assistant to conduct the research. He will have to do most of the work himself assisted by the laboratory technologist.

The major costs in this proposal comes from costs of equipment. Equipment costs contribute to a major portion of the funds requested for in most projects.

Table 2.1: Budget estimates for Project 1.

I. Manpower	
Technologist	$56,600
II. Equipment	
4°C freezer	$4,000
Cryogenic freezer	$40,000
Core reamer set	$8,000
Power saw and driver	$8,000
Cannulated small screw fragment set	$8,000
Subtotal	*$68,000*
III. Materials and Supplies	
Animals (Cost and Maintenance)	$18,740
Consumables	$2,800
Others	$1,500
Subtotal	*$23,040*
IV. Miscellaneous Items	
Nil	$0
Total	**$147,640**

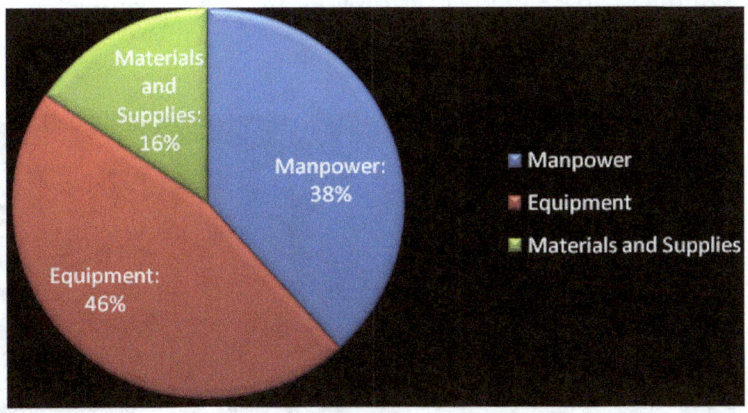

Fig. 2.10: Overview of budget estimate for Project 1.

In this case, note that animal costs amount to only 12.7% of total project costs. This is because only a small animal is used — in this case the rabbit.

In this application, the costs for consumables set aside is too little — only $2,800. It should also include separate costing for anaesthetic items, surgical items, histological items, X-ray costs, photography costs. This should amount to at least $10,000 as will be better shown in the costing of the second project.

The second illustration shows the actual costs in the budget estimates for a Project 2: 'Role of BMP in enhancing anterior lumbar interbody fusion using allografts' spanning a two-year duration, as depicted in Table 2.2 and Fig. 2.11.

It can be seen that in the second proposal, manpower costs amounted to $109,700. This is almost double the amount required for Project 1 ($56,600) even though Project 2 is a year shorter in duration than Project 1. This is because a research assistant is requested for and not just a technologist. Likewise, animal costs ($61,350) are three times that in Project 1. This is because the monkeys used are very expensive compared to the smaller animals (rabbits) used in Project 1. OP-1 (osteogenic protein 1) also took an equally large share of the budget ($61,517) because BMPs are very expensive.

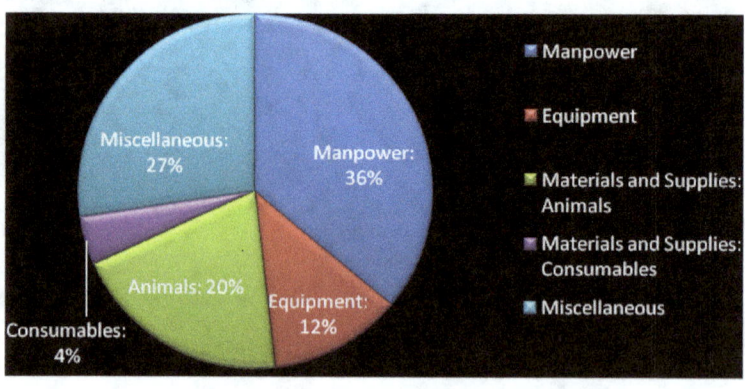

Fig. 2.11: Overview of budget estimate for Project 2.

Table 2.2: Budget estimates for Project 2.

I. Manpower

Research assistant	$109,700

II. Equipment

Diathermy machine	$16,000
Basic orthopaedic set	$6,000
Spinal instrument set	$12,000
Osteotome instrument set	$3,000
Subtotal	*$37,000*

III. Materials and Supplies

Animals

Cost of 24 monkeys + 20% wastage	$1,740 × 30 = $52,200
Maintenance costs ($20 per week per animal + 20% wastage)	
8 weeks (8 monkeys)	10 × $20 × 8 = $1,600
12 weeks (8 monkeys)	10 × $20 × 12 = $2,400
16 weeks (8 monkeys)	10 × $20 × 16 = $3,200
Anaesthesia costs ($50 per animal)	$50 × 30 = $1,500
Disposal costs ($15 per animal)	$15 × 30 = $450
Subtotal	*$61,350*

Consumables

Anaesthetic items	$2,400
Surgical items	$3,000
Histological items	$4,000
Plain X-rays	$2,000
Photography	$2,000
Subtotal	*$13,400*

IV. Miscellaneous Items

X-rays in the Department of Radiology ($100 per specimen)	$100 × 30 = $3,000
CT scans in the Department of Radiology ($500 per specimen)	$500 × 30 = $15,000
Gamma irradiation of lyophilised femoral cortical ring allografts	$4,000
OP-1 (osteogenic protein 1) for 16 vials ($3,204 per vial + 20% wastage)	$3,204 × 16 × 1.2 = $61,517
Subtotal	*$83,517*
Total	**$304,967**

Table 2.3 shows that for Project 2, all the equipment should be purchased in the first year of the project. About 50% of Materials and Supplies as well as Miscellaneous Items must be made available in Year 1 of this two-year project.

Table 2.3: Yearly breakdown of budget estimates for Project 2.

Category	Year 1	Year 2
Manpower	$53,300	$56,400
Equipment	$37,000	
Materials and Supplies	$40,000	$34,750
Miscellaneous	$50,000	$33,517
Total	$180,300	$124,667

Other support

The PI should also indicate whether he is in receipt of funds or other forms of support. For example, the donation of −80°C electrical freezers by a private foundation will put the investigator into a good position in an application for a research project on allograft transplantation.

Likewise, collaborations with industry will be a bonus point in the application of research grant. For example, a certain well-established company may be prepared to co-share the costs of research on the fusion of certain cages for spinal fusion they are interested in. Such collaborations will be favourably looked upon by the review committee provided the product or equipment being researched on has great clinical value potentially. In such cases, correspondences from such companies — or better still contracts — should be included in the appendix.

Biographical Data of Applicants

The application must include biographical data of the PI and his collaborators.

Each sketch should include the academic degrees and the research and professional experience of each applicant in a chronological sequence, concluding with the present position. It should also include a list of selected publications in chronological order, especially those most pertinent to the current research application.

An important part of the decision to approve a project by the research committee will also be based on whether the applicants have the necessary expertise to complete the research. This part of the evaluation is based on the biographical data of all the applicants appended.

Suggested Names of Reviewers

In some institutions, the applicants are required to suggest at least the names of two reviewers. In such cases, the PI is also asked to indicate two reviewers to whom he thinks the application should not be sent for review.

Internal and External Reviews

For research applications not exceeding about $200,000, only internal reviews need to be conducted. However, for research applications exceeding $200,000, external reviews are required.

Chapter 3

Types of Research: An Overview

Jamie Xiang Lee Kee, Haitong Mao & Aziz Nather

Types of Research

One must decide on the type of research one is interested in. Clinical research is the most popular type of research amongst clinicians. A few embark on basic science research, which is more challenging and time consuming. It is important to encourage a core of young clinicians to take the challenge and embark on a basic research project. Only then can a department develop expertise and strength in basic research. The department may need to strengthen and develop the facilities supporting basic research, such as manpower, equipment and expertise which are often lacking. Such facilities need to be slowly acquired.

Clinical Research

Clinical research is research that directly involves a particular person or group of people.[1] It studies the effect of various modes of treatment for patients, e.g., the outcome of new drugs on a patient, or the outcome of a new surgical technique.

Ethical conduct of clinical research

The cardinal rule in medicine is "to do no harm" (Fig. 3.1). As such, to responsibly perform clinical research:

- Patients consenting to be in the research protocol must be informed of the possible outcome (Fig. 3.2).
- Potential risks should be assessed thoroughly, weighed against potential benefits, and minimised.
- Confidentiality of patient information is mandatory.
- Honesty in the reporting of findings is imperative.

Reliability of research

- Proper controls should be designed.
- Randomised trials can be carried out.
- Double-blind clinical trials could also be performed to minimize bias. One may, for example, investigate the effect of Drug X on the relief of symptoms in patients with osteoarthritis of the knee (Fig. 3.3). The doctor prescribing the drug in cream or tablet form does not know which one is the placebo and which one is Drug X.

Fig. 3.1: The hippocratic oath.

Source: http://americantl.org/hippocartic-oath-serpent-staff

Types of Research: An Overview 59

Fig. 3.2: Patient consent must be obtained before conducting any clinical research.

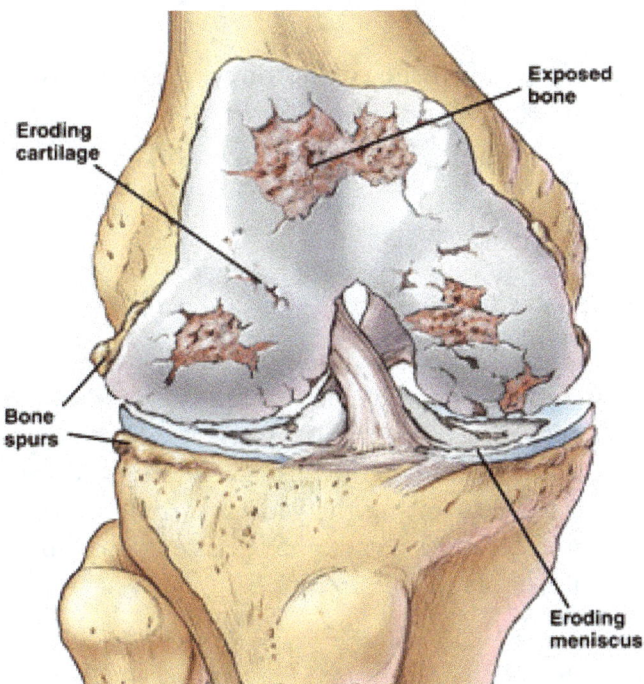

Fig. 3.3: Osteoarthritis.

The code is recorded for each patient. The code is only broken upon completion of the trial after all evaluation has been done by an independent party.

Basic Science Research

Basic science research may include research on biology or biomechanics of healing of the following tissues:

- Skin
- Fat
- Bone
- Cartilage
- Muscle
- Tendon
- Nerves
- Vessels

Another big area is tissue engineering, which involves:

- Stem cells
- Growth factors
- Scaffolds

Cadaveric Research

Good basic research can be conducted using cadavers (Fig. 3.4). An example is a comparison of anterior lumbar interbody reconstruction and fixation using Kaneda Instrumentation with reconstruction using a Z-plate or other anterior instrumentation devices. This is a biomechanical study involving the cadaveric spine. Another example of a biomechanical study is comparing the strength of undreamed nail versus external fixators in fracture of tibia.

Prior considerations

- Fresh cadavers should be used in cadaveric research.
- Ethical codes and the law must be obeyed.

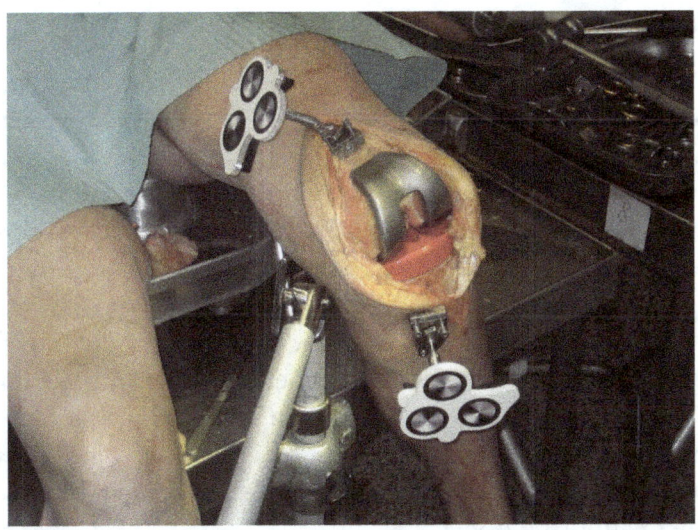

Fig. 3.4:   Research on a knee prosthesis in a cadaver.

- Consent must obtained from:
  - The Department Head of Forensic Pathology of the hospital from which the cadavers are procured
  - Relatives of the deceased subjects
- It is difficult to obtain sufficient cadavers for the project, as the study population must be large enough to be statistically significant.
- The study population must be homogenous in terms of their demographic, e.g., in age, sex and race.

Obtaining the cadaver

Once consent is obtained from the Department of Forensic Pathology, a team must be assembled, ready to perform the harvesting whenever a cadaver is available. The team should include at least a resident to sign for receipt of cadaveric parts. The resident must know how to harvest the cadaveric part without jeopardising the experiment. He must be assisted by at least one or two technologists. One should also work out the logistics of harvesting, such as the necessary surgical instruments and containers to receive the spare parts, and transport arrangements.

Consideration of facilities

- Good facilities, such as proper surgical instrumentation, special equipment and proper jigs for mounting the specimens, must be available.
- Upon harvesting, the spare parts must be stored in a reliable freezer at a temperature of −40°C. There must thus be enough space to accommodate the freezer as well as a dissecting table for the cadaveric dissection.
- To perform biomechanical testing, the Department must have a Shimadzu Universal Testing Machine or an Instrom Machine. One should also collaborate with a bioengineer. Departments lacking such facilities and expertise should engage engineers in a Department of Mechanical Engineering nearby. If not, such biomechanical studies cannot be done.

Animal Experimentation Research

To perform animal experimentation, the Department must first have an Animal Holding Unit. The Unit supplies the necessary animals, provides housing facilities for the animals, provides the necessary anaesthesia and looks after the post-operative care of such animals. It is preferable to have a veterinary surgeon on hand, especially for the anaesthesia of larger animals for experimental surgery.

Choosing your animal

Choosing the correct experimental animal is of paramount importance. One must not simply use the first animal available in the laboratory.

Factors to consider include:

- The type of tissue to be experimented on
- The type of animals used by other workers
- The facilities available in the Animal Holding Unit

Research areas

Research areas worthy of attention include healing of fractures, bone transplantation, healing of tendons and ligaments, healing of nerves, spinal cord regeneration, allograft transplantation and tissue engineering.

Ethics in animal research

It must be remembered that we live in an animal-loving society. While restrictions against animal experimentation in Singapore are not as strict as those in some countries like the USA, this situation must not be abused. One must always bear in mind ethical considerations. Publicity must also be avoided at all costs. As researchers, we should not entertain requests from reporters no matter how tempting the situation is.

Designing Proper Controls

Many make the common mistake of not designing proper controls for the study. The planning of proper controls for each research project is vital. Many experiments become meaningless because of a lack of proper control studies.

In many clinical areas, normal values (controls) are unknown. For example, a study on hindfoot valgus in a rheumatoid foot was rejected by a prominent journal because the research worker did not study the angle of inclination of the hindfoot in normal patients, which was unknown. The researchers thus had to study this before their work on the rheumatoid valgus could be accepted.

The same applies to animal experiments. In the author's study on the healing of bone transplants (autografts) in the tibia of adult cats (Fig. 3.5),[2] there were no data as to how long fractures take to heal in the adult cats. He had no choice but first to design a simple osteotomy model to determine this basic data which is required.

Paired Controls

The best experimental model to use is paired controls. The control animal for each experiment should be selected to be as similar as possible to the tested animal. Paired controls are especially useful in experiments on the extremities of animals. An example would be in the investigation of the healing of vascularised versus non-vascularised bone autograft transplants in adult cats.[2] A good paired control model is to perform vascularised bone transplantation in the right

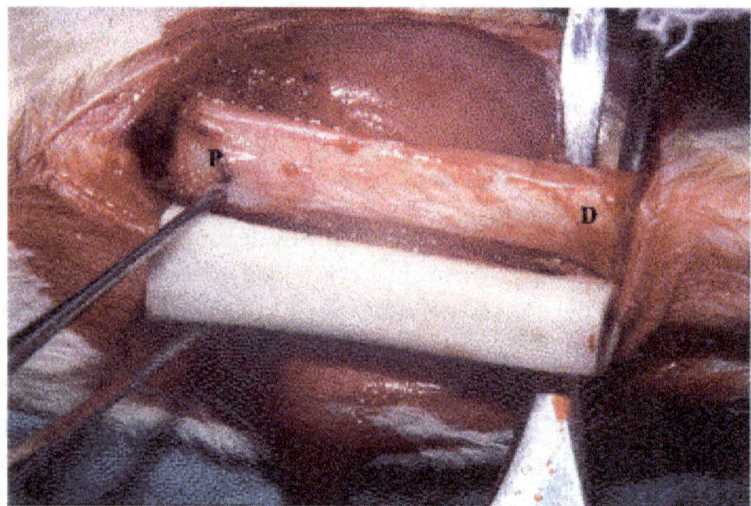

Fig. 3.5: Allograft experiment in adult cats. A photograph showing a 4-cm segment of tibia excised, proximal osteotomy (P) and distal osteotomy (D) performed by an oscillating saw, the allograft to be transplanted (devoid of periosteum) being shown below the segment.

tibia and compare this with non-vascularised bone transplantation in the left tibia of the same cat.

Sham Operations

A sham operation is a placebo operation that omits the procedure thought to be essential for healing. Sham operations must be performed for proper clinical significance. For example, in studying the effect of dividing the anterior cruciate ligament (ACL) in a rabbit, one should perform the whole operation including division of the ligament in the right knee(Fig. 3.6).[3] On the left knee of the same rabbit, a sham operation is performed. This comprises an incision of the skin, subcutaneous tissue and medial parapatellar incision without proceeding further to divide the anterior cruciate ligament.

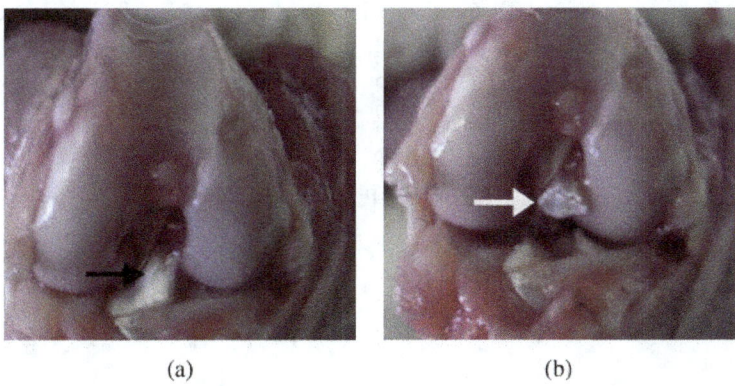

(a) (b)

Fig. 3.6: Operative photos of a rabbit knee before (a) and after (b) transection of the ACL.

Estimation of Projected Budget

Proper budget estimation must be done so the research grant applied for will be able to cover all costs to be incurred in the research project. The budget estimates must include the following items:

- Manpower costs
- Cost of equipment
  - E.g., special equipment, surgical instruments, X-rays, photography

- Cost of materials and supplies
  - E.g., animals, consumables
- Miscellaneous costs

It is extremely important for the researcher to ask for all items necessary to conduct the research. Many research projects are uncompleted because budget estimation is inaccurate. When planning is poor, actual budget costs will exceed the projected budget costs. In such cases, the researcher has no choice but to apply for supplementary funds in order to salvage and complete the research project.

References

1. Eunice Kennedy Shriver National Institute of Child Health and Human Development. (2012). Clinical trials & clinical research, Retrieved from http://www.nichd.nih.gov/health/clinicalresearch/Pages/index.aspx
2. Nather, A., Balasubramaniam, P. & Bose, K. (1988). Revascularisation and fracture healing in a large avascular segment of bone. An experimental study. *J. Bone Joint Surg.* **72B**: 830–834.
3. Satkunanantham, K., Kumar V.P. & Nather, A. (1992). Deterioration following meniscectomy — an exaggeration. *J. Asean Ortho. Assn.* **6**: 27–29.

Chapter 4

Clinical Research

Aziz Nather, Jamie Xiang Lee Kee & Haitong Mao

There are many different types of studies that can be conducted in clinical research. The research field depends on the clinical area of interest of the researcher. Once the specific problem is identified, you should decide the type of study to be conducted. It is important to understand which suits the purpose of your study.

Types of Clinical Studies

Most studies can be split into two major classes: retrospective studies and prospective studies. (Fig. 4.1).

Prospective Studies

Prospective studies are studies in which the question is set prior to data collection. After the study is designed, subjects are identified based on relevant criteria. They are then followed to see if the outcome of interest ultimately occurs as a result of the intervention given.

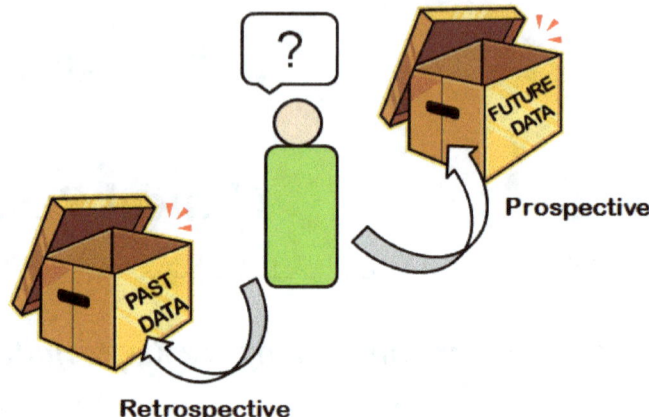

Fig. 4.1: Pictorial depiction of the two different types of clinical studies.

Retrospective Studies

Unlike prospective studies, the occurrence of the outcome of interest in retrospective studies has already been determined before the initiation of the study. Data are usually collected prior to question selection. The data used have not been specifically compiled for the research study, but rather for administration purposes.

> ### Example of a retrospective study
>
> *The investigation of the mortality of elderly patients undergoing hemi-arthroplasty for fracture neck of femur*
>
> The researcher can look up the last 200 consecutive patients operated in the Department over the last two or three years. The medical records and X-ray records of all patients are then traced. Using a detailed research proforma, all data are documented. The patients are recalled for a clinical review to establish the latest outcome of the operation. The study can be completed within a year if it is well planned.

Table 4.1 shows a comparison of prospective and retrospective studies.

Table 4.1: Prospective vs. retrospective studies.

	Prospective	Retrospective
Level of evidence	Evidence obtained is more reliable	Evidence not as strong as that from prospective studies. Unforeseen factors affecting the results may not be taken into account. Nevertheless, a hypothesis can be derived from a retrospective study. The hypothesis can then be further tested.
Time frame	At least 1 to 2 years for the proper implementation of the research protocol. At least 2 to 3 years to complete research study	1 year is a reasonable period of time to complete the study
Execution	More expensive and time-consuming to design and carry out	Relatively easier to perform
Suitable studies	For studies that involve reaching conclusions about the effectiveness of interventions, prospective studies are the most definitive.	As information is already collected, studies of diseases which require a long follow-up period, in which there is a long wait between the exposure and outcome, are better suited to retrospective studies.

Clinical Study Designs

There are a myriad of different study designs with differing levels of evidence. The following will be discussed here:

- Clinical trials (*single-centre trials and multicentre trials*)
- Randomised controlled trials (RCTs)
- Observational studies
- Cohort studies (*prospective, retrospective*)
- Case-control studies
- Case reports and case series

Clinical Trials

Clinical trials are an integral part of prospective studies. Through trials, information is gained about an experimental treatment. Table 4.2 describes the various phases of a clinical trial.

Considerations in planning clinical trials

- Eligibility requirements — The choice of study population depends on the type of trial. If the trial is exploratory, a relatively homogeneous patient population would be preferred. For a management trial, a more heterogeneous population will be able to explore the effect of the therapy in different groups.
- Ethics — Informed consent, etc.
- Response to failure — A plan should be in place to tackle the scenario in which the therapy fails and proves detrimental.

Table 4.2: Phases of a clinical trial.

Phase	Description
1	• To assess the tolerability of new drug through adverse events, clinical and laboratory parameters and toxicology • Determine a safe dosage range, and the effects of food/drug–drug interaction
2	• Proof of concept, dose-range finding or definitive dose-response studies • To determine the minimum and maximum effective or tolerated dose
3	• To confirm effectiveness • To monitor side effects • To compare against commonly used treatments
4	• Usually for post marketing safety surveillance • Sometimes conducted to provide additional details on the drug's long-term safety, efficacy and cost effectiveness • To determine new dosage forms or formulations

Adapted from: http://www.sgh.com.sg/research/listofclinicaltrials/pages/faqsonclinicaltrials.aspx

Single-Centre vs. Multicentre Trials

Single-centre trials

Single-centre trials are clinical trials set up in a single hospital or clinic, while multicentre trials are those carried out in several different medical centres or clinics. Single-centre trials are performed on a smaller scale compared to their multicentre counterparts. They are cheaper to fund and therefore easier to procure funding for. Single-centre trials are usually the starting point from which new treatments sprout, as they offer scientists and clinicians flexibility. Pilot studies and Phase 1 trials are usually single-centre trials. One such example is the study 'Influence of vitamin D status and vitamin D_3 supplementation on genome wide expression of white blood cells: a randomized double-blind clinical trial' conducted by the Boston University School of Medicine,[1] in which they sought to determine the effect of vitamin D on the gene expression in healthy adults.

However, most single-centre trials involve a considerably smaller number of patients than multicentre trials. The smaller sample size thus runs a higher risk of false positives or false negatives (Type I or Type II errors). The results reaped from single-centre trials are thereby harder to generalise than those from multicentre trials. A single centre may also be unable to retain enough consenting patients for the clinical trial to be feasible.

Multicentre trials

Multicentre trials are the most commonly accepted way of evaluating new drugs and technologies. The usage of multiple centres allows a larger number of subjects to be recruited, thus ensuring that the study population will be wide enough to be statistically significant. This increases the validity of the project. Testing the new technology in multiple centres also allows a better basis for the generalisation of the findings as a broader range of clinical settings are explored. Different perspectives on the value of the new technology can also be obtained. Phase 3 trials are usually carried out through multiple centres to obtain better feedback on the tested therapy.

However, multicentre trials are complex, expensive and require a high amount of coordination among the different centres. An efficient control centre overseeing of all the trial activities is required. The manner in which the protocol is implemented must be clear and similar across all centres involved for results to be meaningful. Procedures should be standardised and a common protocol should be implemented. The investigators should meet regularly, all personnel should be trained and the trial should be carefully monitored. Good trial management is pivotal. If possible, a trial manager should be hired. This manager will be tasked with the coordination of the trials across the various centres. Figure 4.2 is a useful checklist for a multicentre trial.

Multicentre Trial Checklist

- All investigators conduct the trial in strict compliance with the protocol agreed to by the sponsor and, if required, by the regulatory authorities.
- The data collection forms are designed to capture the required data at all multicentre trial sites. For those investigators who are collecting additional data, supplemental data collection forms that are designed to capture the additional data should also be provided.
- The responsibilities of coordinating investigator(s) and the other participating investigators are documented prior to the start of the trial.
- All investigators are given instructions on following the protocol, on complying with a uniform set of standards for the assessment of clinical and laboratory findings, and on completing the data collection forms.
- Communication between investigators is facilitated.

Fig. 4.2: The multicentre trial checklist.

Adapted from: International Conference on Harmonisation Guidelines for Good Clinical Practice (ICH GCP), http://ichgcp.net/5-sponsor

Randomised Controlled Trials

Randomised controlled trials (RCTs) are considered the gold standard of study design, and can prove cause-and-effect relationships. They are prospective studies. In a RCT, each patient is assigned randomly to one of two or more treatment groups. The control group is that which receives standard care, no intervention or a placebo. Outcomes are then evaluated after the treatment. Researchers do not know which treatment is superior and any of the treatments chosen could be beneficial to participants.[2]

For instance, in the investigation of the benefits of external fixators versus unreamed intramedullary nailing for open fractures of the tibia, consecutive patients could be assigned to either operation without surgeon bias by using a random system of allocation.

In certain studies, the treatment received by the patient is kept unknown. These studies are referred to as "blinded" studies. There are two types of blinded studies:

- *Single-blind* — Patients do not know which of the treatments they are receiving.
- *Double-blind* — Both patients and their doctors are not aware which of the treatments the patient is receiving. However, should the patient's condition become life-threatening, the doctor can request to un-blind the treatment to better manage the patient's situation.

Example of a double-blind trial

The investigation of the effect of Drug X on the relief of symptoms in patients with osteoarthritis of the knee

The doctor prescribing the drug in cream or tablet form does not know which one is the placebo and which one is Drug X. The code is recorded for each patient. The code is only broken upon completion of the trial after all evaluation has been done by an independent party.

Observational Studies

In an observational study, investigators draw inferences about the possible effects of a treatment whose administration to subjects is outside of the investigator's control. Differences in one characteristic (e.g., whether subjects received a specific intervention) are studied in relation to differences in other characteristics, without the intervention of the investigator.[3] Observational studies are often carried out when a randomised experiment would be impractical, unfeasible or unethical.

Example of an observational study

The investigation of the effect of smoking on lung capacity in women

The investigator finds 100 30-year old women. Of these women, 50 have been smoking a pack of cigarettes daily for 10 years, while the other 50 have been smoke-free for the same duration. The lung capacities of these 100 women are measured. Data obtained is then analysed, interpreted and conclusions are drawn.

In the above example, the "treatment" is not introduced by the investigator, but by the wholly autonomous decision of the subjects.

Observational studies are more likely to result in selection bias. Also, a challenge in observational studies is ensuring that the factors investigated are the causes of recorded effects. Sometimes, important causal factors are not recorded.

In 2007, several prominent medical researchers issued the *Strengthening the Reporting of Observational Studies in Epidemiology* (STROBE) statement, in which they called for observational studies to conform to 22 criteria that would make their conclusions easier to understand and generalise.[3]

Cohort Studies

A cohort study is a type of observational study. It is usually used to analyse risk factors for a disease, and describe the incidence and natural history of a condition.[4] In a cohort study, a sample group, or the cohort, is first identified. A cohort is a group of people who have a common experience or characteristic, e.g., they smoke or are born in the same year. All individuals in the group must have the potential to develop the condition of interest. If one is looking at the natural history of a condition, the sample group must also be representative of the population that one is studying.

A cohort study can be either prospective or retrospective.

Prospective cohort studies

Individuals who do not have the disease or condition of interest are chosen to form the sample group. One then measures and records variables that might lead to the development of the outcome of interest. The members of the sample group are then followed up overtime to observe if they develop the condition of interest.

Two types of controls may be used in such studies. Internal controls are used in single-cohort studies, and they comprise individuals in the same cohort who do not develop the condition of interest. External controls are used when the study involves two cohorts. One cohort has been exposed to the variable that is being studied while the other has not. The unexposed cohort will serve as an external control.[5]

One problem that may be encountered in a prospective cohort study is the inability to follow up with some subjects. This could result in a less accurate result analysis. To avoid this, one should record all contact information of the patient at the start of the study. One should also contact the patient regularly to ensure that he does not lose interest in participating in the study.

Retrospective cohort studies

A retrospective cohort study utilises data that are already available. The data used have been previously collected for some other purpose. Retrospective cohort studies are shorter and cheaper than prospective cohort studies, since one only has to collate and analyse the data.

One advantage a retrospective cohort study has over a prospective one is the lack of data collection bias as the condition of interest was not the reason for the collection of data. However, important and relevant details may have been left out as the data collected were originally meant for some other purpose. Recall bias may also be present, as people with the condition of interest may exaggerate or minimise what they think are risk factors.

Advantages of cohort studies

Cohort studies can be useful in instances where RCTs are not feasible or ethical. One cannot, for instance, deliberately expose subjects to risk factors. A cohort study must be conducted instead.

Cohort studies allow one to distinguish whether a variable is a cause or effect of the condition of interest, as the possible causes are observed before the condition of interest developed.

Additionally, in a cohort study, more than one outcome can be studied at once. For instance, a cohort study of alcoholics can simultaneously look at deaths from liver disease and heart disease. This is unlike case-control studies which only look at one outcome at one time.

A cohort study also allows the importance of each variable, or the relative risk, to be statistically calculated. This calculation is more reliable when the incidence of the outcome is high. The greater the number of subjects in the cohort and the higher the probability of developing the condition of interest, the more effective the cohort study will be.

Disadvantages of cohort studies

The evidence from cohort studies is not as strong due to the presence of confounding variables. Confounding variables are factors that may differ between two subjects leading to different outcomes, aside from the variable that is being studied. Asthma patients, for instance, may have a lower incidence of cardiovascular disease, but this might be due to the fact that asthma patients are less likely to smoke. Smoking here is a confounding variable.

Confounding variables can only be eliminated in a prospective RCT. In such a trial, whether or not the subject is exposed to the risk factor is assigned by chance, and there would be an equal proportion of confounding factors in the exposed and unexposed groups.

In addition, there is also a potential for bias in any cohort study. The subjects chosen for the cohort may not be representative of the population that one is studying. For instance, if the cohort consists of only employed people, the study will be biased as the employed generally enjoy better health than the unemployed.

Some useful tips

If the data are readily available then a retrospective design is the quickest method. If high quality, reliable data are not available a prospective study will be required.

Each variable studied must be accurately measured. Variables that are relatively fixed, such as height, need only be recorded once. Where change is more probable, for example, drug misuse or weight, repeated measurements will be required.

To minimise the potential for missing a confounding variable all probable relevant variables should be measured. If this is not done the study conclusions can be readily criticised. All patients entered into the study should also be followed up for the duration of the study. Beware, follow up is usually easier in people who have been exposed to the agent of interest and this may lead to bias.

Source: Mann, C. J. (2003). Observational research methods. Research design II: cohort, cross sectional, and case-control studies. Emerg. Med. J. 20: 54–60.

Example of a cohort study

Investigating the effect of smoking on lung cancer

The investigator recruits a group of smokers and a group of non-smokers (the unexposed group). The groups are matched as much as possible in terms of many other variables such as economic status and other health statuses to minimise confounding factors. He then follows them for a set period of time and observes the differences in the incidence of lung cancer between the groups.

Case-Control Studies

Case-control studies are considered the most reliable retrospective study because they approximate a control group. They are observational studies as no intervention is attempted. Information is collected on two groups of patients: those with the medical condition of interest, and those without. The former is the case group, while the latter is the control group. Both groups should be as similar as possible in all other ways. The researchers will then retrospectively compare the degree to which the subjects of each group were exposed to a certain variable, such as a treatment or risk factor. The relationship between the variable and condition can thus be determined.

Example of a case-control study

The investigation of the relationship between use of conjugated estrogens and the risk of endometrial cancer

Study population: 188 white women aged 40 to 80 years old with newly diagnosed endometrial cancer and 428 controls of similar age hospitalised for non-malignant conditions required surgery at the Boston Hospital for Women Parkway Division, Massachusetts, between January 1970 and June 1975.

Data on drug use and reproductive variables were extracted from hospital charts and from the medical records of each woman's private physician. 39% of the cases and 20% of the controls had used conjugated estrogens in the past (Buring *et al.*, 1986). The higher prevalence of use of conjugated estrogens among the cases, as compared to the controls, suggested that use of the drug is was associated with an increase in the incidence of endometrial cancer.

Adapted from: dos Santos Silva, I. (1999). *Cancer Epidemiology: Principles and Methods. International Agency for Research on Cancer*, Chapter 9.

Case control studies allow one to look simultaneously at multiple risk factors and are useful as initial studies to establish relationships between variables and conditions. Also, less time is needed to conduct case control studies because the condition has already occurred.

However, they are retrospective and rely on memory — this may potentially lead to recall bias. Care should also be taken to avoid confounding variables (i.e., when an exposure and an outcome are both strongly associated with a third variable).[6] Controls should also preferably come from the same population from which cases are derived to reduce the chance that other differences between the groups account for the difference in the exposure that is under investigation.[7]

Case Reports and Case Series

Case reports and case series are descriptive studies. While a case report usually describes a single patient, a case series is a collection of reports (Fig. 4.3).

Case reports

A case report is normally the report of something that has happened or has been observed, that is, the report is retrospective. Case reports normally focus on the manifestations, clinical course and prognosis or outcome for the patient.[8]

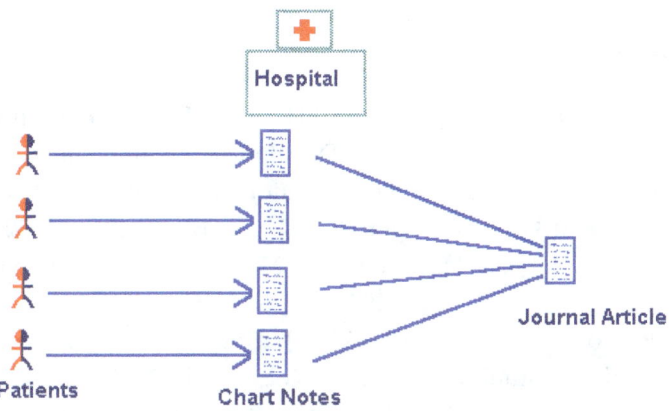

Fig. 4.3: Case series and case reports.

Source: Suny Downstate Medical Center (Medical Research Library of Brooklyn). (2004). Case series and case reports. http://library.downstate.edu/EBM2/2600.htm

Case series

A case series is a descriptive study that consists of a collection of reports on the treatment of patients with the same symptoms, disease discrepancies or outcomes.[8] They can be either prospective or retrospective. They usually involve only a small group of subjects. Based on detailed clinical evaluations and patient histories, an observed effect (i.e., the illness) and a specific exposure may be able to be associated.[8]

However, case series may be undermined by bias since case series report on data from a small subject group without a comparison, or "control", population. This lack of control groups for the comparison of patient outcomes cause case series to be less statistically valid (low-level evidence) than other studies. Additionally, since the researcher is at liberty to select the cases included in the series, selection bias may arise.[9]

Case series are useful when a disease is uncommon or when a disease is caused almost exclusively by a single kind of exposure (e.g., vinyl chloride and angiosarcoma). Case series or case reports can help to offer information that can assist in a diagnosis.[10] Case reports (or case series) may be first to provide clues in identifying a new disease or adverse health effects from an exposure.[7]

References

1. Hossein-nezhad, A., Spira, A. & Holick, M. F. (2013). Influence of vitamin D status and vitamin D_3 supplementation on genome wide expression of white blood cells: a randomized double-blind clinical trial. *PLoS ONE* **8**(3): e58725, doi:10.1371/journal.pone.0058725
2. Moore, A., McQuay, H., Derry, S. & Moore, M. (2013). Randomisation (or random allocation). *Bandolier J.* [Glossary].
3. Vanderbroucke, J. P. *et al.* (2007). Strengthening the reporting of observational studies in epidemiology (STROBE): explanation and elaboration. *PLoS Med.* **4**(10): e297, 10.1371/journal.pmed.0040297
4. Pennsylvania State University. (2013). Lesson 9: Cohort study design; sample size and power considerations for epidemiologic studies. Retrieved from https://onlinecourses.science.psu.edu/stat507/node/54

5. Mann, C. J. (2003). Observational research methods. Research design II: cohort, cross sectional, and case-control studies. *Emerg. Med. J.* **20**: 54–60.
6. The George Washington University (Himmelfarb Health Sciences Library). (2011). Study design 101: case control study. Retrieved from http://himmelfarb.gwu.edu/tutorials/studydesign101/casecontrols.html
7. California Department of Public Health (2013). What is a case-control study? Retrieved from http://www.ehib.org/faq.jsp?faq_key=34
8. Flinders University (2012). Evidence based practice: case report and case series. Retrieved from http://www.caresearch.com.au/Caresearch/Portals/0/Documents/PROFESSIONAL-GROUPS/Nurses%20Hub/NH_EBP_CaseReports_May2012.pdf
9. Suny Downstate Medical Center (Medical Research Library of Brooklyn). (2004). Case series and case reports. Retrieved from http://library.downstate.edu/EBM2/2600.htm

Chapter **5**

Choice of Experimental Animals

Aziz Nather, Jane Jia Xin Lim, & Elaine Yi Ling Tay

Introduction

Choosing the appropriate animal for experimentation is vital to the success of the research project. It is a cardinal sin to choose the first animal available in the experimental laboratory.

Literature Review

A literature search must first be performed on animal experimental models used by previous research workers in the same field.

When the investigator only uses intermediate animals such as artiodactyls, e.g., pigs (*Sus scrofa*) or mountain goats (*Oreamnos americanus*), the results would be less significant as compared to other work involving primates, e.g., monkeys *(Macaca fascicularis)*.

In contrast, when the researcher chooses primates, the results produced would be more significant compared to results of other work using small animals such as lagomorphs, e.g., rabbits (*Oryctolagus cuniculus*) or rodents, e.g., rats *(Rattus norvegicus)*.

The higher the position of the animal chosen on the phylogenetic order of animals, the greater the significance of the results attained. Experiments using primates and intermediate animals would be more significant than those employing small animals.

One must also anticipate problems encountered with different animal models. Authors tend to downplay the difficulties they experience. It is useful to consult a veterinary surgeon to discuss problems that may arise.

The investigator must also seek the use of an animal holding facility that can not only indent the animals needed, but also provide anaesthesia, operating surgery facility, post-operative analgesia and antibiotic prophylaxis as well as day-to-day housing, care and maintenance of the animal experimented upon.

Phylogenetic Classification of Animals

Animals are classified according to the Linnaean system.[1] Table 5.1 shows the taxonomic classification of common animals used.

Primates are closest to man (*Homo sapiens*) in terms of evolutionary development (Table 5.1). Experimental studies using primates such as monkeys (*Macaca mulatta*; Fig. 5.1a), or their cousins (*Macaca fascicularis*; Fig. 5.1b) would be more relevant in clinical application to man than those using non-primates.

This relationship is illustrated in a research paper titled 'Monkeying around with HIV vaccines: Using rhesus macaques to define gatekeepers for clinical trials'.[2] In the study of human disease and the development of vaccines against HIV and AIDS, the author used rhesus macaques as the animal model because of the close relationship of monkeys to man in the phylogenetic order of animals.

In contrast, result of studies using small animals, e.g., randomly bred rats which are much lower in the evolutionary scale such as rats (*Rattus norvegicus*), may not be so clinically relevant for application to man.

Table 5.1: Taxonomic classification of common animals used.

Common name	Class	Order	Family	Genus	Species
Man	Mammalia	Primate	Hominidae	*Homo*	*sapiens*
Monkey	Mammalia	Primate	Cercopithecidae	*Macaca*	*mulatta* (rhesus) or *fascicularis* (crab-eating)
Baboon	Mammalia	Primate	Cercopithecidae	*Papio*	*hamadryas*
Pig	Mammalia	Artiodactyla	Suidae	*Sus*	*scrofa*
Goat	Mammalia	Artiodactyla	Bovidae	*Oreamnos* or *Capra*	*americanus* (mountain) or *hircus* (domestic)
Sheep	Mammalia	Artiodactyla	Bovidae	*Ovis*	*aries*
Dog	Mammalia	Carnivora	Canidae	*Canis*	*lupus familiaris*
Cat	Mammalia	Carnivora	Felidae	*Felis*	*catus* (domestic) or *silvestris* (wild)
Rabbit	Mammalia	Lagomorpha	Leporidae	*Oryctolagus*	*cuniculus*
Guinea Pig	Mammalia	Rodentia	Caviidae	*Cavia*	*porcellus*
Rat	Mammalia	Rodentia	Muridae	*Rattus*	*rattus* (black) or *norvegicus* (brown)
Mouse	Mammalia	Rodentia	Muridae	*Mus*	*musculus* (house)

In this regard, if primates (Fig. 5.2) cannot be procured, it is far better to use intermediate animals such as arteriodactyles, e.g., pigs (*Sus scrofa*) and carnivores, e.g., dogs (*Canis lupus familiaris*) or cats (*Felis silvestvis*) than to use small animals such as lagomorphs, e.g.,

Fig. 5.1a: Macaca mulatta.
Source: http://www.ecologyasia.com/verts/mammals/rhesus-macaque.htm

Fig.5.1b: Macaca fascicularis.
Source: http://www.redbubble.com/people/liefranky/works/4939111-macaca-fascicularis

rabbits (*Oryctolagus cuniculus*) (Fig. 5.2). Artriodactyles and carnivores are higher in evolution than lagomorphs. The latter is higher than rodents. All three are lower than primates.

While dogs are larger than cats, both are equal in terms of evolutionary development. They both belong to the same order Carnivora (Fig. 5.2) Therefore, no significant advantage is obtained in experiments involving dogs compared to those using cats.

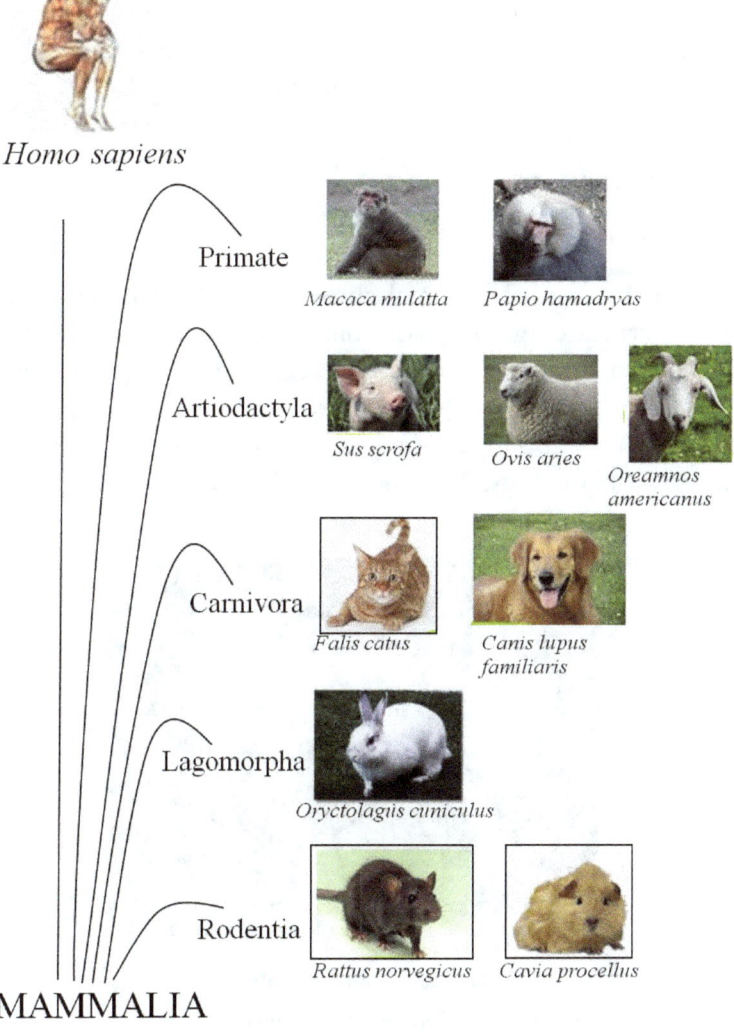

Fig. 5.2: **Evolutionary tree of Mammalia.**

Experimental Models

Certain tissues are well developed in some animals compared to others, e.g., ligaments of the knee and ankle are better developed in animals that jump such as rabbits (Fig. 5.3) than in "non-jumpers".

Likewise, the brain in the cat is higher in evolutionary development than that in dogs and other animals. Depending on the tissue to be experimented upon, e.g., tendon, a particular animal known to house well-developed tendon should be chosen.

Tendon Healing Model

Ligaments in the knee

For tendon healing, the animal chosen must be a jumper. To study healing of cruciate ligaments or collateral ligaments in the knee, the animal chosen must have large hind legs and a fairly well-developed knee joint, e.g., the rabbit (*Oryctolagus cuniculus*; Fig. 5.3). The knee ligaments in the rabbit are larger in size compared to those in carnivores, e.g., cats or dogs. The knee joint in the cat or dog is small.

Fig. 5.3: Jumping rabbit.
Source: http://anythingpet.blogspot.sg/2007/10/rabbit-jumping.html

Achilles tendon

To study the repair of tendons, the anatomical location of the tendon selected and the size of the tendon must be considered before choosing the experimental model. The Achilles tendon in the rabbit is a fairly large tendon to study healing of different techniques of tendon lengthening. Nather *et al.*[3] studied three different techniques of tendon-lengthening (Z-lenthening, intramuscular lengthening and tenotomy) using the flexor digitorum longus of the hindleg in the white rabbit.

Tendons in the hand

In contrast, to study the healing and function of tendons in the upper limb, particularly the tendons in the hand, primates would be the most appropriate experimental model (Fig. 5.4). Chacha *et al.*[4] found the flexor tendon mechanism in macaque monkeys to be remarkably identical to those in man. Even the arrangement of the vinculae was similar. This was also observed by Kaplan[5] in gorillas and chimpanzees. He studied the transplantation of autologous composite tissue tendon grafts with intact fibrous flexor sheath procured from the third toe (longest toe) into the middle, ring and little fingers in seven macaque monkeys.

Fig. 5.4: Hand of a primate.
Source: http://www.foundshit.com/monkey-blocking-face/

Hyaline Cartilage Model

Choice of joint

The knee joint would be more suitable than the hip joint in designing an experimental model on hyaline cartilage of a weight-bearing joint. This is because the knee joint is superficial and readily accessible while the hip joint is deep-seated. The ankle joint is much smaller in size even though it is superficial.

Choice of animal

Having chosen the knee joint, the more appropriate animal to use is a rabbit rather than a cat or a dog. This is because the knee joint is better developed in rabbits. Satku *et al.*[6] studied the effect of bilateral meniscectomy alone and bilateral meniscectomy plus division of the anterior cruciate ligament in New Zealand white rabbits.

Osteoarthritis model

Bilateral meniscectomy plus sectioning the anterior cruciate was found to produce osteoarthritis of the knee.[6]

Continuous passive motion of the knee

Robert Salter championed the hypothesis that continuous passive motion stimulated pluripotent mesenchymal cells to differentiate into articular cartilage.[7] He performed a series of experiments using the knee joint of adolescent New Zealand white rabbits. One study demonstrated the chondrogenic potential of free periosteal graft in a synovial fluid environment. Another study also demonstrated the stimulation effect of continuous passive motion on periosteal neochondrogenesis.[8] In another experimental model, he showed good results demonstrating the chondrogenic potential of autogenous osteoperiosteal grafts to repair major osteochondral defects by using continuous passive motion.[9]

However, despite a series of experiments performed, Salter's results of using continuous passive motion (CPM) remained controversial. Several research workers felt that since the experimental model used was the adolescent New Zealand white adult rabbit rather than mature adult rabbits, the results demonstrated could only perhaps be clinically relevant to children and not to adults. He has not reproduced such results in experiments employing adult rabbits. This uncertainty could be removed if the experiments were to be repeated using mature rabbits (epiphyses shown to be fused on radiographs) or using an intermediate animal model, e.g., adult pig (*Sus scrofa*).

Nevertheless, despite the controversy, Salter collaborated with Saringer, an engineer to develop CPM devices for humans (Fig. 5.5).[7]

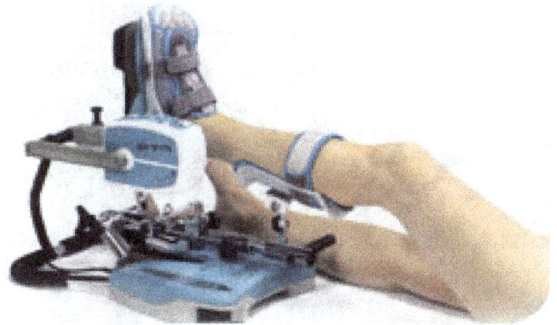

Fig. 5.5: CPM treatment using motor-driven exercise equipment.
Source: http://www.htherapy.co.za/Continuous_Passive_Motion

Fracture Repair Model

Man has a remarkable capacity to heal fractures. Indeed, the ability of higher vertebrates to heal bone is good and equals that of lower vertebrates. In man, fracture repair occurs by proliferation of fibrovascular and osteogenic tissue without going through an early cartilaginous phase.[10] Some cartilage may be present in the callus. The single most important factor is the stability of the fracture. Movement promotes chondrogenic maturation and callus proliferation. The amount of callus and cartilage around unstable fractures is usually greater than around those fractures well stabilised.[11]

Small animals

Fractures in small animals unite more quickly. Many experiments in rats and mice show that large amounts of cartilaginous periostal callus are produced (early cartilaginous phase). This is later transformed into bone (later osteogenic phase) through endochondral ossification[10] (Fig. 5.6a). However, the proportion of cartilage in the dog and man is usually less than that present in the rodent (Fig. 5.6b).

For these reasons, the rodent is not favored as an experimental animal for fracture repair. The rat is rarely used.[12] Rabbits have been

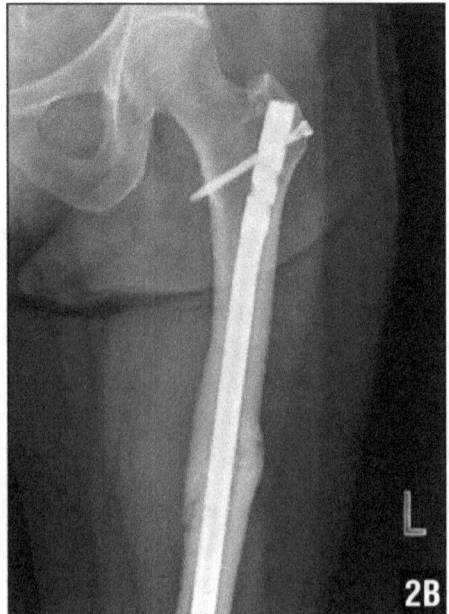

Fig. 5.6a: Fracture healing in human.

Adapted from: http://www.healio.com/orthopedics/journals/ortho/%7Bcc08e0c6-0122-4688-b165-9e1f00d92fd3%7D/atypical-diaphyseal-femur-fractures-in-patients-with-prolonged-administration-of-bisphosphonate-medication-for-osteoporosis

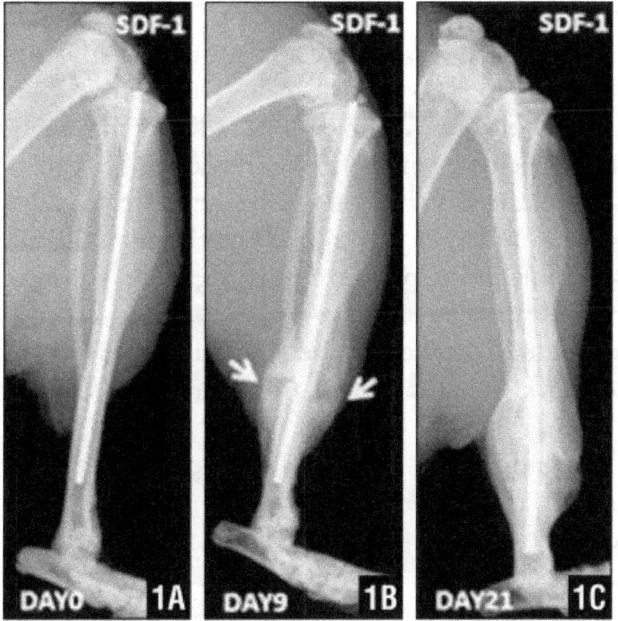

Fig. 5.6b: Fracture healing in mouse.

Adapted from: http://www.healio.com/orthopedics/journals/ortho/%7B676e06c4-9e26-43b3-8598-40239aeaf3db%7D/single-percutaneous-injection-of-stromal-cell-derived-factor-1-induces-bone-repair-in-mouse-closed-tibial-fracture-model

used,[11,13] but they are less suitable compared to carnivores. The popular animal chosen is the dog.[14–20] Other animals used include the cat, the sheep and rarely the monkey.

Bone Transplantation Model

The experimental animal chosen for bone transplantation is usually the dog[21–28] or the rabbit.[29,30] Occasionally, the monkey[31] or the rat[32] is employed. A review of the literature in English reveals only one study using adult cats.[33]

A popular model for studies on cortical bone transplant involved excising a 4-cm segment of the fibula in adult dogs.[24,28,34] However, comparative osteology of the tibia and fibula in the dog, cat and rabbit (Figs. 5.7a and 5.7b) showed that in comparison to the bones in the dog or the rabbit, the tibia and fibula in the adult cat were similar in morphology to that of man. They remain as two separate bones in the whole leg.[35] In the dog, radiographs taken (Fig. 5.7b) showed that the entire length of the fibula can be seen. The fibula is separated from the tibia[12,36] in the proximal part although in the lower half the fibula is closely applied to the tibia. In the rabbit, the fibula is seen only in the proximal portion, being fused to the tibia in the lower half of the leg (Fig. 5.7b).

Nather therefore preferred the feline tibial model[37] to the canine fibular model for bone transplantation studies for several reasons:

1. The fibula in the dog is not a fully weight-bearing bone since it is joined to the tibia in its lower half.
2. A 4 cm segment in the dog's fibula is a relatively small bone transplant (less than one-third of the dog's leg). On the other

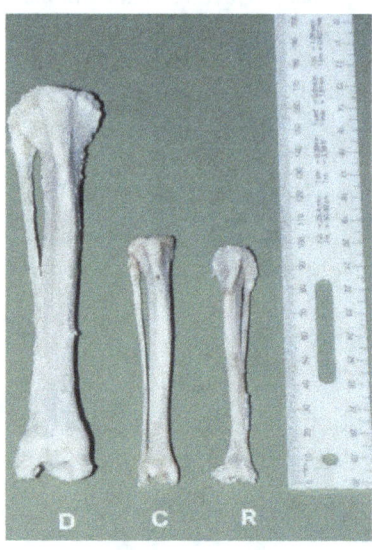

Fig. 5.7a: Comparative osteology of tibia and fibula in dog, cat and rabbit. In the dog (D) and in the rabbit (R), the tibia and fibula are morphologically fused in the lower half of the leg. In the cat (C), the two bones remain as separate bones throughout the whole length.

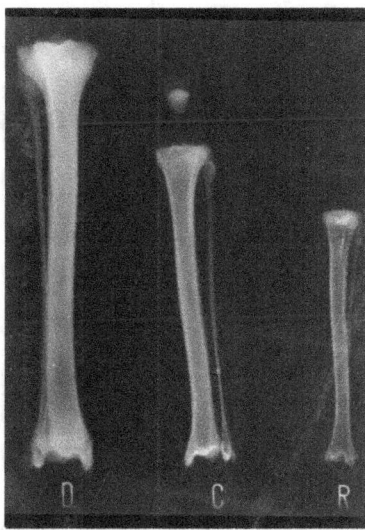

Fig. 5.7b: Radiologically in the dog (D), the entire length of the fibula can be seen, although in the lower half the fibula is partially fused to the tibia. In the rabbit (R), the fibula is fused to the tibia in the lower half of the leg. In the cat (C), the tibia and fibula remains as separate bones throughout.

hand, a 4 cm segment in the cat's tibia is proportionately a large bone transplant (at least two-thirds of the tibial diaphysis).
3. There is no need to perform internal fixation in the feline tibial model. The intact separate fibula acts as an internal splint. Cast immobilisation of the cat's lower limb is adequate for immobilisation, thereby eliminating one additional variable in the experiment, namely the need for internal fixation using a metal implant.

Nather used this experimental model successfully to study the biology and biomechanics of healing of both non-vascularised autografts[11] and allografts (Figs. 5.8a–d).

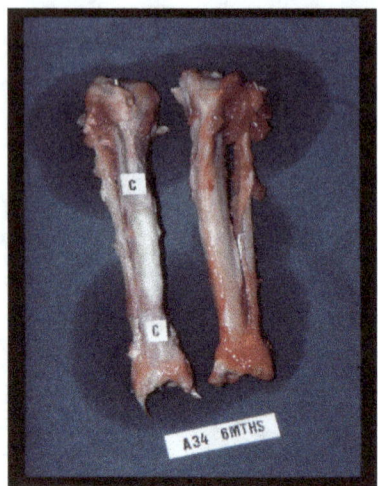

Fig. 5.8a: A 4-cm segment of tibia excised (A) and replaced with identical segment of tibial allograft (B).

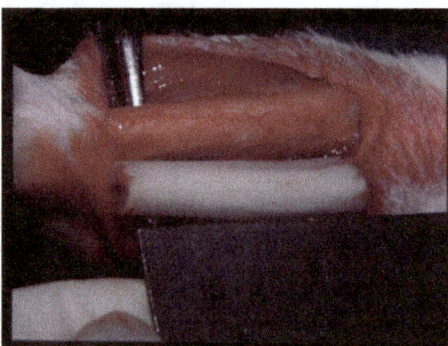

Fig. 5.8b: A six-month specimen of allograft showing union at both junctions in the right tibia.

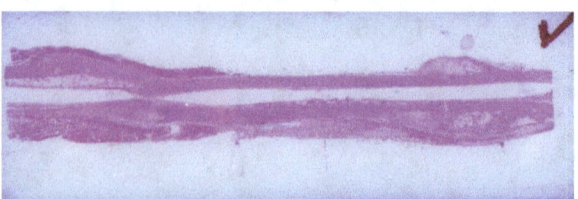

Fig. 5.8c: Histological section of a 16-week specimen showing osseous callus at both junctions.

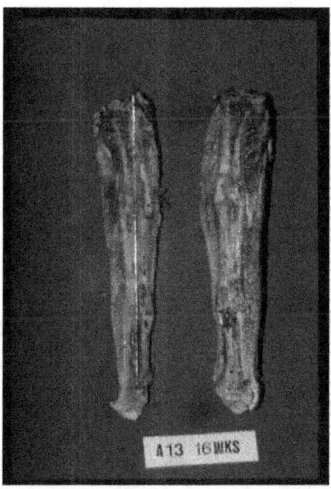

Fig. 5.8d: A 16-week allograft specimen bisected into half to show an intramedullary rod and union at both junctions.

Spinal Interbody Fusion Model

For research on spinal interbody fusion, one must look for an animal with a spine large enough to allow the operation to be performed easily. A primate (*Macaca fasicularis*) is ideal. However, if this is not available, the animal model chosen should at least be an intermediate animal, e.g. a dog, pig, sheep or mountain goat.

Vertebral formula

The vertebral formula in the pig (*Sus scrofa*) is $C_7T_{14-15}L_{6-7}S_4Cy_{20-23}$[12], the formula in the goat (*Oreamnos americanus*) is $C_7T_{13}L_6S_5Cy_{12}$ whilst the formula in the sheep (*Ovis aries*) is $C_7T_{13}L_{6-7}S_4Cy_{16-18}$. Excluding the cervical region, variation in number is common.[12] (Sission, 1975)

The dog (Fig. 5.9a) and goat (Fig. 5.9b) have been used because the size and anatomy of the spine is similar to that of man. Small animals such as rabbits and rats are not suitable choices because their spines are too small to work upon.

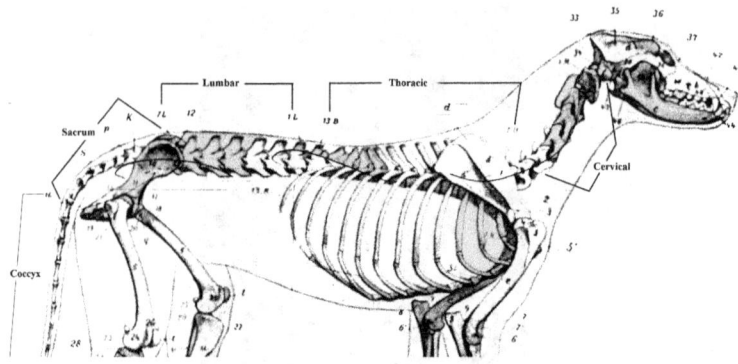

Fig. 5.9a: Spine of the dog with the vertebral formula $C_7T_{13}L_7S_3Cy_{20-23}$.

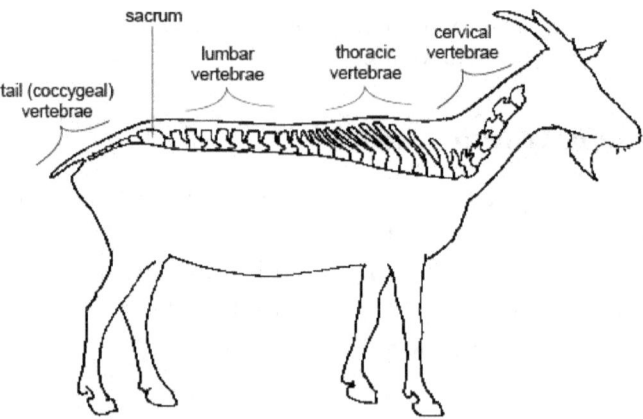

Fig. 5.9b: Spine of the goat with the vertebral formula $C_7T_{13}L_6S_5Cy_{12}$.

Source: http://commons.wikimedia.org/wiki/File:Anatomy_and_physiology_of_animals_Regions_of_a_vertebral_column.jpg

Nerve Repair Model

Unfortunately, very few studies have been done on healing of nerves. Chacha *et al.*[38] studied experimental sensory reinnervation of the median nerve by nerve transfer using *Macaca fascicularis* as the experimental animal.

Facilities Available for Animal Holding

Before a research worker can conduct an animal experimental study, he must first gain access to a facility for animal holding. A good facility indents for the animals required. It also houses the animals in suitable cages and provides food and care for the animals. It also provides anaesthetic facilities and operating room facilities to perform the operations required including good post-operative analgesia and prophylaxis against infections.

Capacity of Animal Holding Facility

The choice of experimental animals must take into account the capacity of the animal holding facility available. It takes into account the type of animal the facility could provide and also the number of animals that could be supported for the research period. (Figs. 5.10a–d). The number of small animals, e.g., rabbits that could be housed at any one time could be large (approximately 20–30), but for larger animals, e.g., pigs and monkeys, this number is more limited (4–5). Smaller animals are also easier to anaesthetise than larger animals, e.g., the pig and sheep. Primates are the most difficult animals to perform experiments on.

A veterinary surgeon should be available to provide expertise for the anaesthesia of animals and also to provide expertise on other aspects of the animal experimented. The veterinarian can advise on whether the type of surgery performed is appropriate, e.g., whether the animal could survive the type of surgery being performed and also give advice on post-operative care of the animal.

NUS Laboratory Animals Centre

The National University of Singapore (NUS) Laboratory Animals Centre (LAC) in Sembawang (Fig. 5.11) has played a pivotal role in the life sciences since 1970.[39] It holds a variety of conventionally bred as well as specific pathogen-free (SPF) animals with a capacity of 10,000 animals. Bred according to internationally recognised breeding systems, the animals are able to remain genetically defined. The LAC supplies about 50,000 laboratory animals annually to institutions such as NUS and Nanyang Technological University (NTU), polytechnics, hospitals and

Fig. 5.10a: Animal-stand parking area. An animal stand holding animal cages in racks mounted on rollers for ease of movement for daily cleaning.

Fig. 5.10b: Rabbit cages arranged in racks using the "book-shelf" principle.

Fig. 5.10c: Cat cage.

Fig. 5.10d: Monkey cage.

 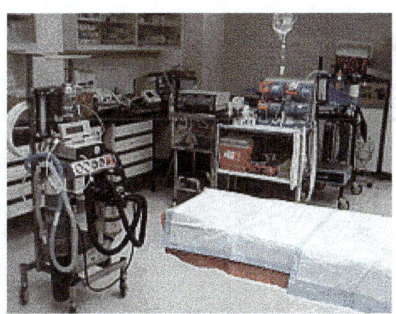

Fig. 5.11: NUS LAC.
Source: http://nuhs2.url3.net/html/pagetree/view_5307.html

other research institutes or centres. Animals are also supplied to institutions in neighbouring countries such as Malaysia, Thailand and Brunei.

NUS Animal Holding Unit

The Animal Holding Unit (AHU) in Singapore is located in the NUS campus.[39] It is an excellent facility in Singapore with a staff of 37 personnel including one full-time veterinary surgeon and a director, who is a consultant from the NUS Department of Microbiology. Apart from holding animals under conventional and isolated conditions, the AHU provides basic laboratory, procedure rooms, X-rays, necropsy, histology, tissue culture and operating suite facilities (Figs. 5.12a–c).

Availability of Animals

In most laboratories, mice, rats, guinea pigs and rabbits are readily available experimental models.

Intermediate Animals

The research worker can use intermediate animals such as pigs (Fig. 5.13a) and sheep (Fig. 5.13b). For this, the researcher has to book an animal holding facility that can support the anaesthesia and surgery on animals, and provide post-operative analgesia and antibiotic prophylaxis. The faculty must also be responsible to indent such large animals.

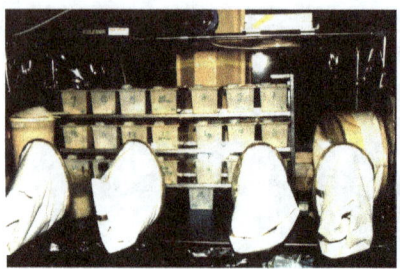

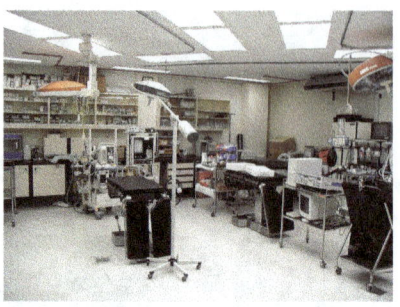

Fig. 5.12a: SPF room with nude mice within a sterile environment. Only filtered air enters the cabinet containing the mice.

Fig. 5.12b: Operating room suite for surgery on animals. The room is lead-lined to allow on-table X-rays to be performed.

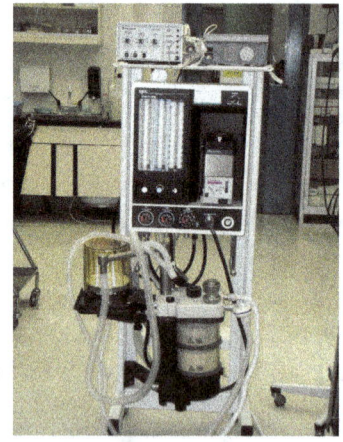

Fig. 5.12c: An anaesthetic machine for inducing sleep in animals.

Primates are available for experimentation in only a few countries (Fig. 5.13c). Monkeys (*Macaca fascicularis*) and baboons (*Papio hamadryas*) are available in Singapore and Indonesia.

Stray Animals

The National University Hospital Department of Orthopaedic Surgery has contracted special veterinary centres for the use of stray dogs and cats. Stray cats have been used by the author for all experimentation

Fig. 5.13a: Pig.
Source: www.all-the-news.com

Fig. 5.13b: Sheep.
Source: www.123rf.com

Fig. 5.13c: Baboon.
Source: www.123rf.com

on non-vascularised autografts and allografts.[39] The stray cats are subjected to quarantine in a separate quarantine bay for at least two weeks. All sick animals are sacrificed and disposed. Only healthy animals that survive after two weeks are chosen for use in the experimental study.

With proper quarantine and maintenance, the wastage rate of such animals is 10% to 20%, which is similar for rabbits.

In summary, the choice of the animal to be selected for the research study depends on the following factors:

- Ability of the facility to indent for the animal requested
- Size of the facility
- Presence of special caging facilities for the animal

- Expertise to handle and anaesthetise the animal
- Manpower running the facility

Procurement of Animals

The Experimental Research Laboratory in the AHU at NUS is a good experimentation facility.[39] The AHU is able to supply larger animals ordered from special abattoirs that meet the necessary requirements of providing pathogen-free animals. It can also procure primates. However, the researcher must give adequate notice to allow time for the faculty to procure the animals requested. Detailed planning between the principal investigator and the AHU must be performed before starting any research project involving the use of animals.

Indentation of Animals

The laboratory animals provided by AHU include:

- Mice
- Random-bred rats (*Rattus norvegicus*) such as Spraque–Dawley (SD) or Wistar (WI) and inbred rats
- Hamsters (*Mesocricetus auratus*)
- Gerbils (*Meriones unguiculatus*)
- Guinea pigs (*Cavia porcellus*)
- Rabbits (*Oryctolagus cuniculus*) such as New Zealand white and local white hybrids

Cost of Animals

The choice of the animal to be used for the experimental study also depends on the budget or research grant available to conduct the research. In budgeting costs of the animals selected for research, the total costs for the animal survey includes:

- Cost of purchase
- Cost of housing

- Cost of feeding
- Cost of maintenance
- Cost of anesthesia, operating time, implants, antibiotics and dressings

Cost of Purchase

Smaller animals mean smaller purchasing costs. For instance, rats and rabbits are relatively inexpensive compared to sheep, pigs and goats (Table 5.2). Primates are very expensive and thus require a larger proportion of the total research budget.

Table 5.2: Cost of experimental animals.[40]

	Type	Cost
Mice	Outbred — ARC(s)	$7.14 (>4 weeks, per animal)
	Inbred — Balb/c, C57BL/6J, CBA/CaH	$10.71 (>4 weeks, per tank)
	F1 Hybrid — Bred on demand	$15.47 (>4 weeks, per cage)
Rats	Outbred: SD & WI	$8.10 (between 4–5 week, per tank)
Guinea Pigs	Hartley	$24.26 (>301–500 g, per tank)
Rabbits	New Zealand White	$106.72 (between 2.5–2.99 kg, per animal)

Cost of Feeding

Smaller animals such as rodents consume less food than larger animals. Special food pellets provide balanced diet for rats and rabbits. Food requirements for larger animals are more costly.

Cost of Housing and Maintenance

Smaller animals such as rats and rabbits are easier to house in small cages. For cats, special cages need to be constructed. Dogs require more space. They are housed in kennels. Primates such as monkeys and baboons are very expensive to maintain. Small animals are cheaper to maintain than larger animals (Table 5.3). The higher maintenance costs for large animals require higher budgeting for animal experiments than using intermediate animals and primates.

Table 5.3: Housing costs of experimental animals at the NUS AHU.

Animal	Cost per diem in 2010*
Mouse	$1.88 per cage
Rat	$2.70 per cage
Guinea pig	$1.80 per animal
Rabbit (Maintenance Feeding — 300 g/day)	$3.63 per animal
Rabbit (Ad-Lib Feeding)	$4.53 per animal
Pig (<45 kg)	$4.84 per animal
Pig (46–59 kg)	$7.26 per animal
Pig (>60 kg)	$12.34 per animal
Goat	$4.84 per animal
Frog	$1.00 per tank

*Non-Inclusive of 7% GST

Duration of Project

The type of animal chosen for research and the number of animals to be experimented upon depends on the duration of time the research project has been allocated. If the research duration is short — 1 to

2 years — only small animals can be chosen and the number of specimens to work on is also limited. More time is required for the study involving primates and intermediate animals such as pigs or sheep to reach an adequate cohort size. It would be difficult to complete an experiment on 12 monkeys in just two years.

Cohort Size

If the cohort size required is more than 24, it is difficult to choose the dog because kennel space is limited. Small animals such as rabbits and rats will be more appropriate to work upon.

Ethical Considerations

In recent years, there has been increased concern on the proper handling and care of animals.

National Bodies

Set up on 12 November 2004, the NUS Institutional Animal Care and Use Committee (IACUC) oversees the care and use of animals for scientific purposes in NUS. Appointment of the members of the IACUC is done via the institutional official. The members include veterinarians, scientists, non-scientists and members of the community who are not affiliated to NUS. The mission of the IACUC is to achieve the best practice by international standards of animal ethics in all animal care and use programmes in NUS.

To achieve its goal, the IACUC evaluates animal use in research, teaching, behavioural and environmental studies through the review of animal use protocols. The IACUC also ensures the housing and care of animals are provided in accordance with the National Advisory Committee for Laboratory Animal Research (NACLAR) Guidelines, titled Guidelines on the Care and Use of Animals for Scientific Purposes (Fig. 5.14). This guideline is implemented in facilities licensed by the Agri-Food and Veterinary Authority of Singapore (AVA), the regulatory body of the Singapore Animals and Birds Act.

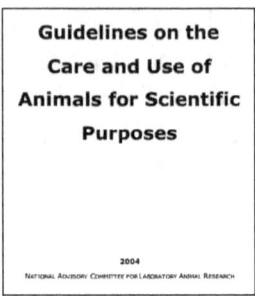

Fig. 5.14: Guidelines on the care and use of animals for scientific purposes.
Source: http://www.ava.gov.sg

The IACUC also ensures that personnel involved in care and use of animals are adequately qualified and trained, through the training programme of the Comparative Medicine (CM). It conducts a semi-annual inspection of animal housing facilities, as well as a semi-annual review of animal care and use programmes. Furthermore, the IACUC also conducts regular inspections of animal use facilities and laboratories to ensure compliance of approved animal use protocols.

International Bodies

The World Organisation (Fig. 5.15) for Animal Health (OIE) is the inter-governmental organisation responsible for improving animal health worldwide.[43] It is recognised as a reference organisation by the World Trade Organization (WTO). The OIE maintains permanent relations with 45 other international and regional organisations and has regional and sub-regional offices on every continent. In 2013, it had a total of 178 member countries.

In the Terrestrial Animal Health Code 2012 published by the OIE, Chapter 7.8 ('Use of Animals in Research and Education') contains international guidelines that set standards to protect the welfare of animals.[41] It is ideal for all research workers to be aware and comply with all these ethical considerations.

Fig. 5.15: Logo of the OIE, world organisation for animal health.

Pilot Study

Before deciding on the type of animal to be selected for the research study, it is essential to first conduct a pilot study to assess the suitability of the animal for the research. Factors to be considered include:

- Availability of the animal
- Anatomy and physiology of the tissue to be studied
- Housing, care and handling facilities available for the animal
- Ease of anaesthesia for the animal
- Ability of the animal to recover from surgery performed to avoid high morbidity and mortality rates

The animal to be adopted for the research must only be confirmed when the pilot study shows the animal to be a suitable experimental model. This will avoid the unsatisfactory situation of changing the experimental animal mid-stream because the researcher was unaware that the animal selected was actually inappropriate for the research model designed.

References

1. Moody, P. A. (1962). *Introduction to Evolution* (2nd ed.). New York: Harper & Row, pp. 308–310.
2. Sherlock, D. J., Silvestri, G. & Weiner, D. B. (2009). Monkeying around with HIV vaccines: using rhesus macaques to define 'gatekeepers' for clinical trials. *Nat. Rev. Immunol.* **9**(10): 717–728.

3. Nather, A., Balasubramaniam, P. & Bose, K. (1986). A comparative study of different methods of tendon lengthening: an experimental study in rabbits. *J. Paed. Ortho.* **6**: 456–459.
4. Chacha, P. B. (1971). Autologous composite tissue tendon grafts for division of both flexor tendons within the digital theca of the fingers (Thesis for Doctor of Medicine). Retrieved from the National University of Singapore.
5. Kaplan, E. B. (1965). *Functional and Surgical Anatomy of the Hand* (2nd ed.). Philadelphia: Lipincott Wilkins & Wilkins, pp. 62–66.
6. Satkunanantham, K., Kumar, V. P. & Nather, A. (1992). Deterioration following meniscectomy — an exaggeration? *J. Asean Ortho. Assn.* **6**: 27–29.
7. Salter, R. B. (1989). The biologic concept of continuous passive motion of synovial joints. The first 18 years of basic research and its clinical application. *Clin. Orthop.* **242**: 12–25.
8. O'Driscoll, S. W. & Salter R. B. (1984). The induction of neochondrogenesis in free intra-articular periosteal autografts under the influence of continuous passive motion. An experimental investigation in the rabbit. *J. Bone Joint Surg.* **66**(A): 1248–1257.
9. O'Driscoll, S. W. & Salter R. B. (1986). The repair of major osteochondral defects in joint surfaces by neochondrogenesis with autogenous osteoperiosteal grafts stimulated by continuous passive motion. An experimental investigation in the rabbit. *Clin. Orthop.* **208**: 131–140.
10. Sevitt, S. (1981). *Bone Repair and Fracture Healing in Men*. Edinburgh: Churchill Livingstone, pp. 306–310.
11. Yamagishi, M. & Yoshimura, Y. (1955). The biomechanics of fracture healing. *J. Bone Joint Surg.* **37**(A): 1035–1068.
12. Wray, J. B. & Lynch, C. J. (1959). The vascular response to fracture of the tibia in the rat. *J. Bone Joint Surg.* **41**(A): 1143–1148.
13. Rahn, B. A., Gallinaro, P., Baltensperger, A. & Perren, S. M. (1971). Primary bone healing: an experimental study in the rabbit. *J. Bone Joint Surg.* **37**(B): 492–505.
14. Rhinelander, F. W. & Baragry, R. A. (1962). Microangiography in bone healing i. undisplaced closed fractures. *J. Bone Joint Surg.* **44**(A), 1273–1298.

15. Rhinelander, F. W., Phillips, R. S., Steel, W. M. & Beer, J. C. (1968). Microangiography in bone healing II. displaced closed fractures. *J. Bone Joint Surg.* **50**(A): 643–662.
16. Rhinelander, F. W. (1974). Tibial blood supply in relation to fracture healing. *Clin. Orthop.* **105**: 34–81.
17. Schenk, R. K. & Willeneger, H. (1963). *Zum histologischen bild der sogenannten primarheilung der knockenkompakta nach experimentellen osteotomen am hund.* *Experimentia* **19**: 593.
18. Richany, S. F., Sprinz, H., Kraner, K., Ashby, J. & Merrill, T. G. (1965). The role of the diaphyseal medulla in the repair and regeneration of the femoral shaft in the adult cat. *J. Bone Joint Surg.* **47**(A): 1565–1584.
19. Nather, A., Balasubramaniam, P. & Bose, K. (1982). Revascularisation and fracture healing in a fracture in a large avascular segment of bone. An experimental study in cats. 7th congress wopa, Perth, Western Australia.
20. Gothman, L. (1961). Arterial changes in experimental fractures of the monkey's tibia with intra-medullary nailing. *Acta Chir. Scand.* **121**: 56–66.
21. Phemister, D. B. (1914). The fate of transplanted bone and regenerative power of its various constituents. *Surg. Gynec. Obstet*, **XIX**: 303–333.
22. Davis, J. B. & Taylor, A. N. (1952). Muscle pedicle bone grafts. An experimental study. *Arch. Surg.* **65**: 330–336.
23. Ostrup, L. T. & Fredrickson, J. M. (1974). Distant transfer of a free, living bone graft by microvascular anastomoses. An experimental study. *Plast. Reconstr. Surg.* **54**: 274–285.
24. Enneking, W. F., Burchardt, H., Paul, J. J. & Pitrowski, G. (1975). Physical and biological aspects of repair in dog cortical — bone transplants. *J. Bone Joint Surg.* **57**(A): 237–252.
25. Doi, K., Tominaga, S. & Shibata, T. (1977). Bone grafts with microvascular anastomoses of vascular pedicles. An experimental study in dogs. *J. Bone Joint Surg.* **59**(A): 809–815.
26. Haw, C. S., McCO'Brien, B. & Kurata, T. (1978). The microsurgical revascularisation of resected segments of tibia in the dog. *J. Bone Joint Surg.* **60**(B): 266–269.
27. Berggren, A., Weiland, A. J. & Ostrup, L. T. (1982). Bone scintigraphy in evaluating the viability of composite bone grafts revascularised by

microvascular anastomoses, Conventional autogenous bone grafts and free non-revascularised periosteal grafts. *J. Bone Joint Surg.* **64**(A): 799–809.

28. Dell, P. C., Burchardt, H. & Glowczewskie, F. P. (1985). A roentgenographic, biomechanical and histological evaluation of vascularised and non-vascularised segmental fibular canine autografts. *J. Bone Joint Surg.* **67**(A): 105–112.
29. Kingma, M. J. & Hampe, J. F. (1964). The behaviour of blood vessels after experimental transplantation of bone. *J. Bone Joint Surg.* **46**(B): 141–150.
30. Medjyesi, S. (1965). Healing of muscle pedicle bone grafts. An experimental study. *Acta Orthop. Scand.* **35**: 294–299.
31. Chacha, P. B., Ahmed, M. & Daruwalla, J. S. (1981). Vascular pedicle graft of the ipsilateral fibula for non-union of the tibia with a large defect. An experimental and clinical study. *J. Bone Joint Surg.* **63**(B): 244–253.
32. Burwell, R. G. (1964). Studies in the transplantation of bone. vii. The fresh composite homograft-autograft of cancellous bone. An analysis of factors leading to osteogenesis in marrow transplants and in marrow-containing bone graft. *J. Bone Joint Surg.* **46**(B): 110–140.
33. Groves, E. W. H. (1917). Methods and results of transplantation of bone in the repair of defects caused by injury or disease. *Br. J. Surg.* **5**: 185–242.
34. Burchadt, H., Busbee, G. A. & Enneking, W. F. (1975). Repair of experimental autologous grafts of cortical bone. *J. Bone Joint Surg.* **57**(A): 814–819.
35. Sisson, S. (1975). Carnivore osteology. Part II Feline. In: *The Anatomy of the Domestic Animals* (5th ed.). Getty, R. (ed.) Philadelphia: WB Saunders Co., pp. 1494–1497.
36. Dekleer, V. S. (1982). Development of bone. In: *Bone in Clinical Orthopaedcis. A Study in Comparative Osteology.* Sumner-Smith, G. (ed.) Philadelphia: WB Saunders Co., pp. 41–46.
37. Nather, A. (1990). The cat as an experimental animal for studies on bone with reference to fracture healing and bone transplantation studies. *J. West Pac. Ortho. Assn.* **27**: 25–32.

38. Chacha, P. B., Krishnamurti A. & Soin K. (1977). Experimental sensory reinnervation of the median nerve by nerve transfer in monkeys. *J. Bone Joint Surg.* **59**(A): 386–390.
39. Nather, A. (2002). *Research Methodology in Orthopaedics and Reconstructive Surgery*. Singapore: World Scientific, pp. 59–65.
40. *Care animal rates.* (2010, August 18). Retrieved from http://www.nus.edu.sg/compmed/services/care-animal-rates.html
41. OIE World Organisation of Animal Health. (2013). Terrestrial animal health code. Retrieved from http://www.oie.int/international-standard-setting/terrestrial-code/access-online

Chapter 6

Cadaveric Research

Elaine Yi Ling Tay, Jane Jia Xi Lim & Aziz Nather

Introduction

Good basic research can be conducted using cadavers. A survey of the *Journal of Bone and Joint Surgery*, *Plastic and Reconstructive Surgery* and the *Journal of Hand Surgery* will reveal that 3 to 5 articles per publication are based on cadaveric dissection. This approach is often thought of as old fashioned and of low technology, especially with the panoply of imaging techniques now available. Though knowledge of anatomy has increased tremendously in recent years, the dissection of cadavers is the simplest and most direct approach to the discovery of the human body. The word anatomy originates from the ancient Greek *ana temnein*, which means to 'cut up' or to 'dissect'.

Prior Considerations

There are some prior considerations before conducting any cadaveric research:

- Fresh cadavers should be used in cadaveric research.
- Ethical codes and the law must be obeyed.

- Consent must be obtained from:
  - The Department Head of Forensic Pathology of the hospital from which the cadavers are procured
  - Relatives of the deceased subjects
- It is difficult to obtain sufficient cadavers for the project, as the study population must be large enough to be statistically significant.
- The study population must be homogenous in terms of their demographic, e.g., in age, sex and race.
- Cadaveric studies take approximately 1 to 1.5 years. The time needed to procure cadaveric parts should be taken into account.

Facilities Available

Freezer Storage Capacity

Cadavers that have been procured are stored in freezers in the mortuary or the cadaveric dissection laboratory. Unlike cadavers used by medical schools for the teaching of anatomy, cadavers used for research are not embalmed before dissection. Upon harvesting, the organs are stored in a freezer under a temperature of $-25°C$ (Fig. 6.1).

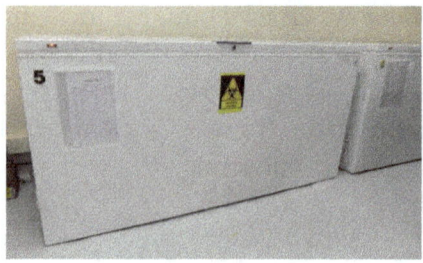
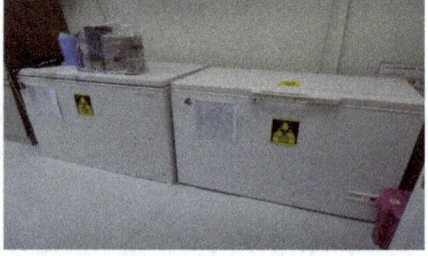

Fig. 6.1: Cadaveric freezers.

Dissection Tables

The dissection tables are properly lined with Incopad to absorb any fluid released during dissection and there must be sufficiently bright lighting at the table (Fig. 6.2).

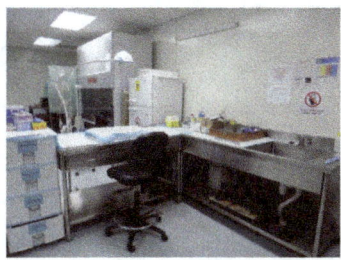 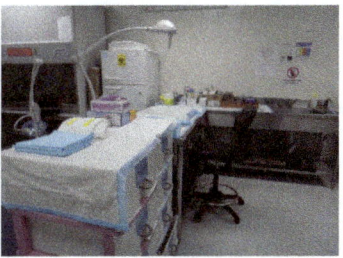

Fig. 6.2: Cadaveric dissection table with Incopad lining and a table lamp.

Instruments for Dissection

Good facilities, such as proper surgical instrumentation, special equipment and proper jigs for mounting the specimens, must be available.

Dissection tools used by the researcher include:

- Skin marker pen
- Scalpel handles and blades
- Disposable scalpels
- Scissors
- Hemostat clamps
- Forceps
- Needle holders
- Spatula probe
- T-pins
- Mallet

The dissection lab would have a wide variety of dissection tools available (Fig. 6.3).

However, the most basic and essential tools required are (Fig. 6.4):

- Scalpel blade holder
- Forceps
- Needle holder
- Scissors

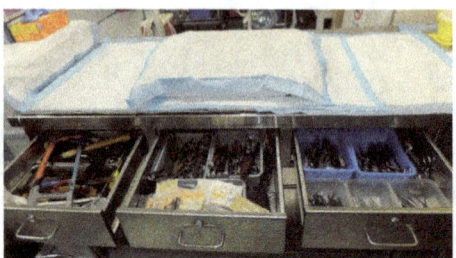

Fig. 6.3:   Dissection tools.

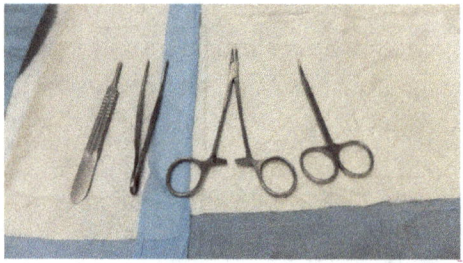

Fig. 6.4:   Basic dissection tools.

Fume Hood

It may be necessary to conduct the dissection in the fume hood (Fig. 6.5). This is to prevent one from inhaling alcohol or toxic formaldehyde.

NUS department of Orthopaedic Surgery Cadaveric Dissection Laboratory

The Cadaveric Dissection Laboratory in the National University of Singapore (NUS)[1] is a service laboratory providing support to a wide range of research that requires validation of a model. The model can be of artificial, biological or artificial–biological construct. The laboratory also handles specimens for cadaveric workshops and courses.

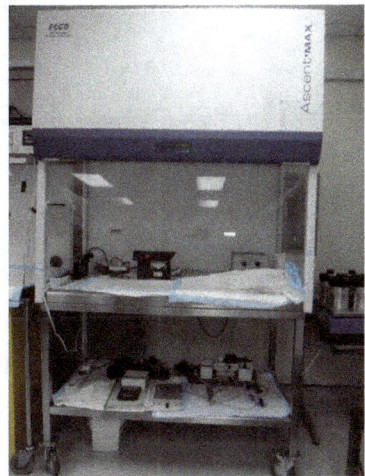

Fig. 6.5: **Fume hood.**

The Cadaveric Dissection Laboratory supports:

- The dissection of cadaveric specimens (Figs. 6.6 and 6.7)
- Preparation of specimens
- Storage and transport of cadaveric specimens
- Thawing and cleaning of specimens before use for both
  - Research work in Orthopaedic Biomechanics Laboratory
  - NUH/NUHS (National University Health System) workshops and courses

Procurement of Cadavers

In Singapore (Fig. 6.8)

The Medical (Therapy, Education and Research) Act in Singapore is an opt-in scheme under the Ministry of Health (MOH), where people of any nationality aged 18 years and above can pledge to donate their organs or any body parts for the purposes of transplant, education or research after they pass away. The Department of Forensic Pathology of the hospital in which the donor has passed away would

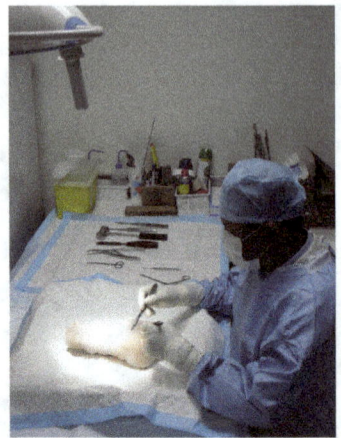

Fig. 6.6: Researcher dissecting cadaveric spine prior to performing biomechanical testing.

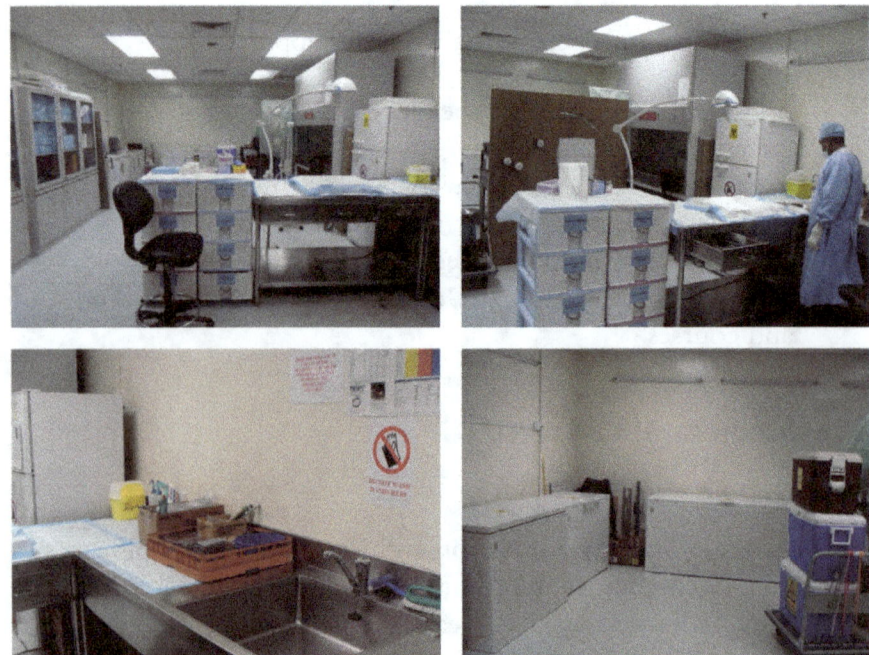

Fig. 6.7: Overview of layout of cadaveric dissection laboratory.

Source: http://medicine.nus.edu.sg/os/facilities/biomechanics/index.html

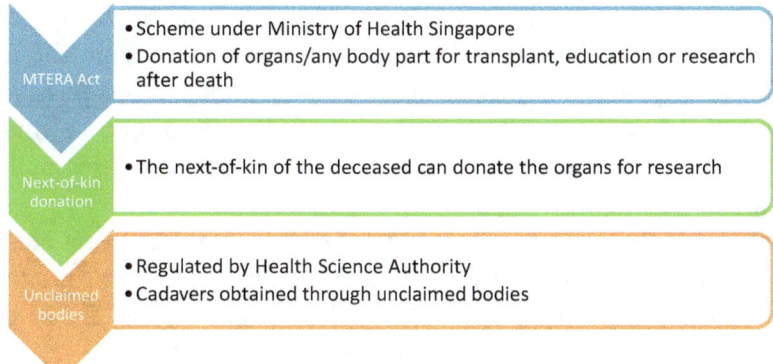

Fig. 6.8: Routes of donation of cadavers in Singapore.

be in charge of the procedure of harvesting the donated organs. Cadavers are also obtained through unclaimed bodies supplied by the Health Sciences Authority (HSA) or donations made by next of kin.

Once consent is obtained from the Department of Forensic Pathology, a team must be assembled, ready to perform the harvesting whenever a cadaver is available. The team should include at least a resident to sign for receipt of cadaveric parts. The resident must know how to harvest the cadaveric part without jeopardising the experiment. He must be assisted by at least one or two technologists. One should also work out the logistics of harvesting, such as the necessary surgical instruments and containers to receive the spare parts, and transport arrangements.

Cadavers from Other Countries

There is a critical supply crunch of cadavers in Singapore.[2] Obtaining sufficient cadavers locally can be very difficult. In order to supplement the shortage, Singapore procures cadavers from overseas, mainly from the USA. These certified institutions from the USA receive the human tissues from voluntary body donations.

When procuring cadaveric tissues, very clear specifications must be given to the institution with regard to the body part wanted as well as the particulars of the donor such as age, sex and weight range. For instance, when procuring a leg, specifications such as from the femur to the toe tips must be indicated. When procuring a shoulder, specifications such as from the clavicle, scapula to the fingertips must be indicated. This is so that all the necessary parts for research are obtained.

In addition, import clearance must be obtained from National Environmental Agency (NEA) (Fig. 6.9). The estimated time period in which the imported parts will arrive in Singapore and the serology report of the cadaver must also be provided.

Fig. 6.9: Logo of NEA Singapore.
Source: http://app2.nea.gov.sg/index.aspx

Preparation Before Dissection

Thawing

Before dissection, the cadaveric parts must be removed from the freezer and thawed for different periods of time depending on the parts.

Table 6.1 summaries the amount of time required for thawing.

Table 6.1: Time required for thawing of cadaveric parts.

Cadaveric body part		Time needed for thawing/days
Head	Without brain	1.5
	With brain	2
Upper torso	Without organs	2.5
	With organs	3
Upper limb		1
Lower limb		1

Safety Guidelines and Rules

Cadaveric dissection laboratory are rated as Bio Safety Level 2 (BSL-2). In addition to the chemicals used to preserve the cadavers such as formaldehyde which is carcinogenic, the dissection kit also contains sharp instruments. Safety guidelines of the lab must thus be strictly followed to ensure the researcher's own safety as well as the safety of others in the lab.

Attire

Researchers must follow an attire code following the Personal Protective Equipment (PPE) requirements when in the laboratory (Fig. 6.10):

- Disposable shoe covers
- Lab coat

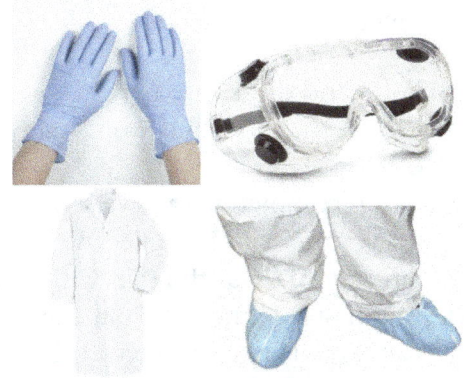

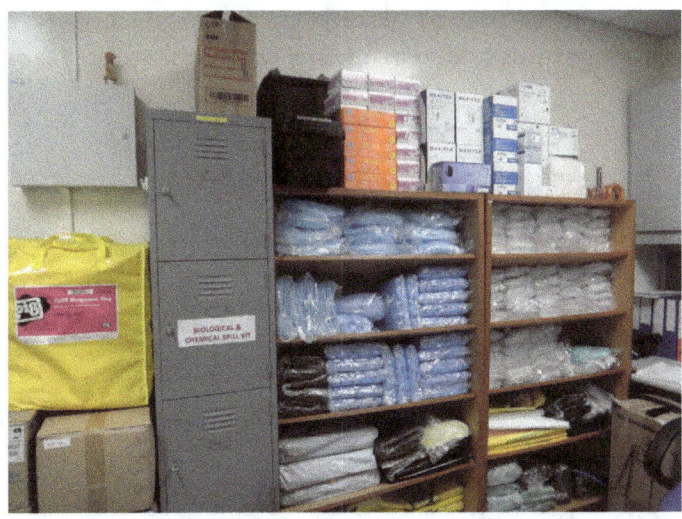

Fig. 6.10: Storage of shoe covers, lab coats, disposable gowns and gloves.

- Scrub gown
- Double gloves
- Safety goggles
- Face shield

Rules

Many rules have to be observed in the laboratory to ensure safety of the researcher as well as the others in the laboratory. The following are some rules that must be observed when conducting research involving cadavers:

- Food and drinks are never allowed in the laboratory.
- Closed toe shoes and long pants or scrubs must be worn to protect the legs.
- It is recommended to wear old clothes, tops with short sleeves or sleeves that can be rolled up with long sleeved lab coats to avoid direct contact of the skin with any fluid.
- Any tissues removed from the cadaver must be placed in a designated container and kept with the body.
- Dispose of all scalpel blades by removing the blade carefully with a hemostat and placing the blade in the sharps container.
- Do not leave tools and instruments on the dissecting table.

Safety Precautions

In case of emergency, the lab should be equipped with the necessary equipment to tackle the situation (Figs. 6.11 and 6.12).

Cadaveric Research 125

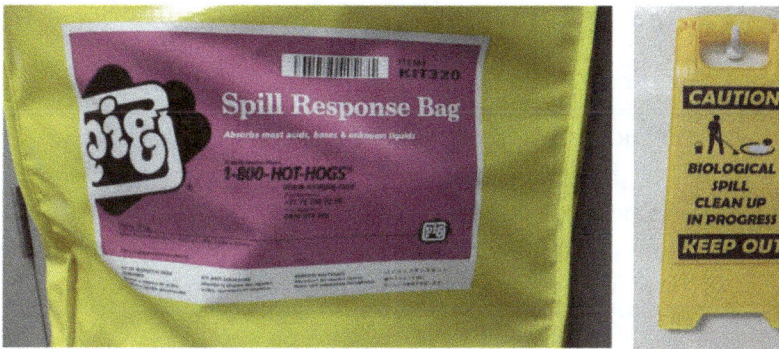

Fig. 6.11: Spill response bag and signage.

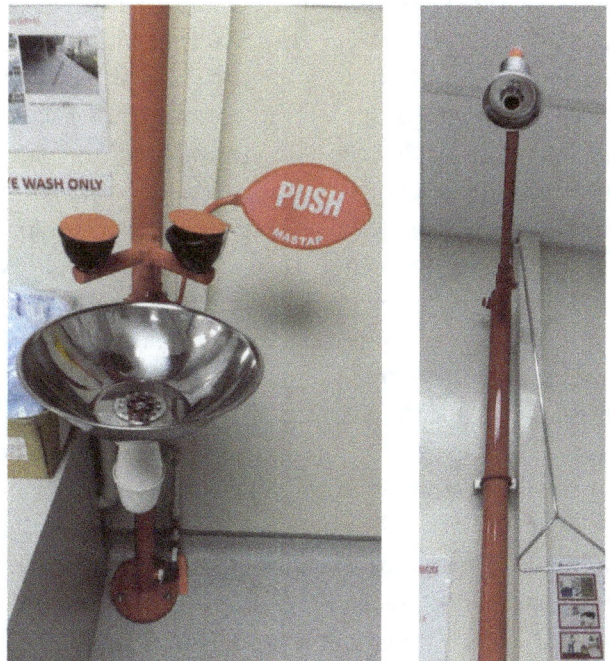

Fig. 6.12: Eyewash and safety shower in the event of a chemical spill.

Areas of Research

Anatomic Research

Criteria for selection of cadavers (such as age, sex, and ethnicity) must be described in detail. Typically, the study population must be as homogenous as possible in terms of age, race and sex, to ensure that valid comparisons can be made. Fresh-frozen cadaveric body parts have to be thawed overnight prior to conducting dissections for research.

Shoulder

Grossman *et al.*[3] previously examined the biomechanical effects of capsular changed in a cadaveric model. It has been speculated that a shift of the throwing arc commonly develops in athletes who perform overhead activities, resulting in greater external rotation and decreased internal rotation cause by anterior capsular laxity and posterior capsular contracture, respectively. Ten cadaveric shoulders were tested with a custom shoulder-testing device. They found that a posterior capsular contracture with decreased internal rotation does not allow the humerus to externally rotate into its normal posteroinferior position in the cocking phase of throwing. Instead, the humeral head is forced posterosuperiorly, which may explain the etiology of Type-II superior labrum anterior-to-posterior lesions in overhead athletes.

McMahon *et al.*[4] developed a novel cadaveric model for anterior–inferior shoulder dislocation using forcible apprehension position. The scapulae of 14 cadaveric entire upper limbs were each rigidly fixed to a custom shoulder-testing device. The cadaveric model yielded an anterior dislocation with a mechanism of forcible apprehension positioning when the appropriate shoulder muscles were simulated and a passive pectoralis major muscle was included. Capsulolabral lesions resulted, similar to those observed *in vivo*.

Hip

Telleria *et al.*[5] examined and described the normal anatomic intra-articular locations of the hip capsular ligaments in the central and peripheral compartments of the hip joint. Eight paired fresh-frozen human cadaveric hips were carefully dissected free of soft tissue to expose the hip capsule. Needles were placed through the capsule along the macroscopic borders of the hip capsular ligaments. Arthroscopy was then performed on each hip. The author found that the hip capsular ligaments have distinct and consistent arthroscopic locations within the hip joint and are associated with clearly identifiable landmarks in the central and peripheral compartments. The standard hip arthroscopy portals are closely related to the borders of the hip capsular ligaments. These findings will help orthopaedic surgeons know which structures are being addressed during arthroscopic surgery and may aid in the development of future hip procedures (Fig. 6.13).

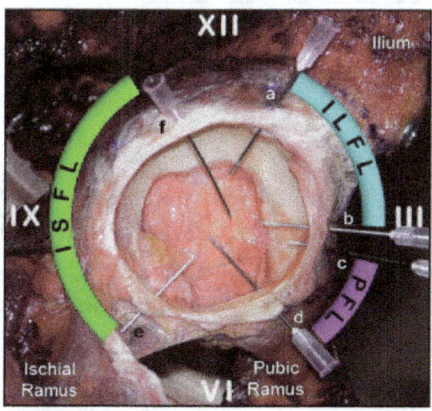

Fig. 6.13: Annotated photograph showing a right acetabulum *in situ* from a cadaveric specimen. The needles demarcate the borders of the hip capsular ligaments. The roman numerals represent hours on a clock face: III o'clock is anterior; VI o'clock is inferior; IX o'clock is posterior; and XII o'clock is lateral. (a, ILFL lateral border; b, ILFL medial border; c, PFL lateral border; d, PFL medial border; e, ISFL inferior/medial border; f, ISFL superior/lateral border.

Source: http://www.sciencedirect.com/science/article/pii/S0749806311000442

Knee (Fig. 6.14)

There is a need to use cadaveric knees because knee kinematics is best examined in actual knees. The anterior cruciate ligament (ACL) is an

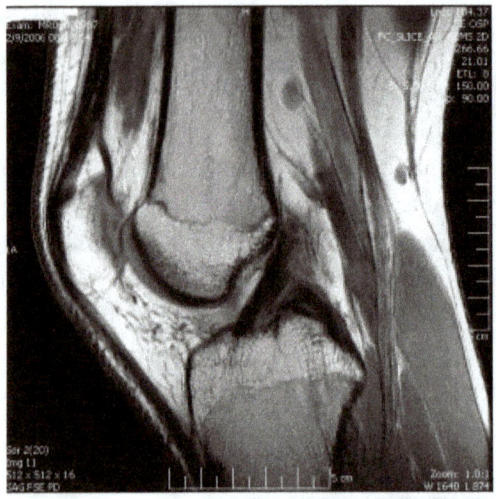 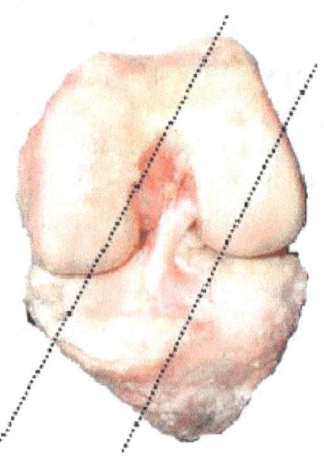

Fig. 6.14: The cadaveric and MRI view of the anterior cruciate ligament (ACL).
Source: http://mme.uwaterloo.ca/~nchandra/?page_id=8

important ligament of the knee. Tan et al.[6] examined anterior cruciate ligament reconstruction in the knees of 30 Singaporean Chinese cadavers (Fig. 6.14). The study revealed that the ACL in Singaporean Chinese is shorter and narrower than those in the Western population. The orientation of ACL is also more vertical. This suggests that placement of the femoral tunnel in ACL reconstruction has to be in a more vertical position to reproduce the physiometry of the ACL.

Ankle and foot (Fig. 6.15)

Cadaver models offer the advantage that there is complete access to all the tissues of the foot, but the cadaver must be manipulated and loaded in a manner which replicates how the foot would have performed when *in vivo*. The cadaver must be mounted on a mechanism that has as many degrees of freedom as the human body. Loads must be applied to the specimen and its tendons at a magnitude and rate as occurs in gait (or as close as possible). Moving the specimen and loading the individual tendon and tibia/foot structures must be

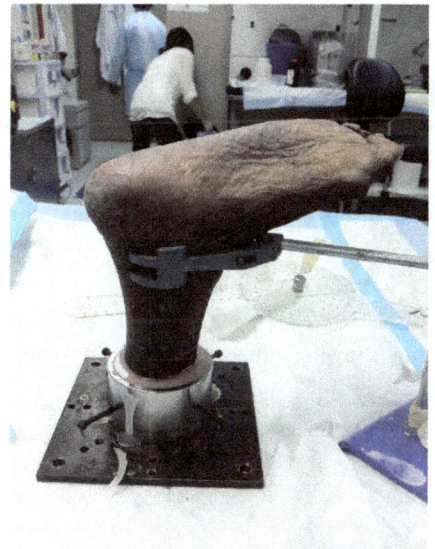 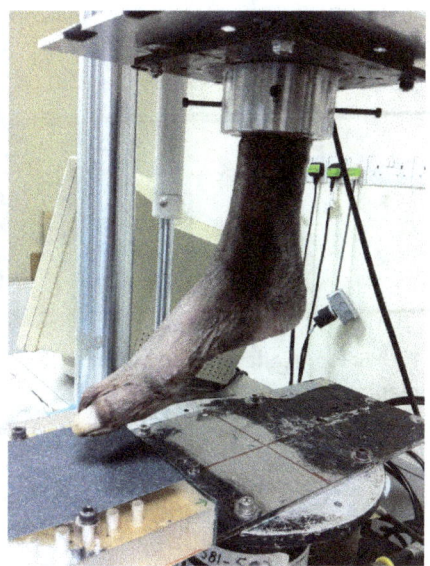

Fig. 6.15: Cadaveric foot and ankle.

synchronised exactly. These parameters must also be adjustable as the input data driving the dynamic model (typically tibial motion, forces applied to the tibia/plantar surface and residual tendons) is at best an average of a small number of other feet, and certainly not *in-vivo* data from the foot being tested.[7]

Elbow and hand (Fig. 6.16)

MacAvoy and Green[8] critically evaluated the current standard for manually assessing strength both clinically and in the scientific literature: the 0-to-5 scale that evolved from the post-World War II report of the British Medical Research Council (MRC). Although widely used by surgeons and researchers, no physical analysis of its validity is available. With the elbow used as an example, a static physics experiment is devised to quantify the enormous difference between maximum strength (grade 5/5) and the strength needed to statically flex against gravity (grade 3/5).

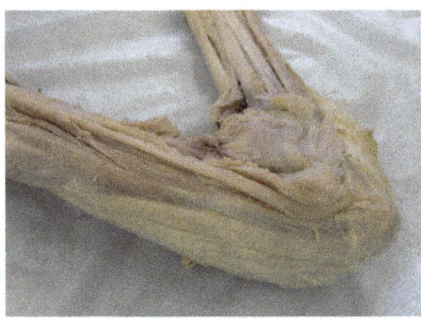

Fig. 6.16: Cadaveric elbow in >90° flexion showing ulnar nerve forced against medial epicondyle.

Source: http://hth.sagepub.com/content/16/3/75/F6.expansion.html

Spine (Figs. 6.17 and 6.18)

Cadaveric spines are carefully dissected from the surrounding soft tissues and muscles to preserve the bone and spinal ligaments prior to testing. Following which, the spines were screened for abnormal anatomy using anteroposterior and lateral fluoroscopy. All spines were stored in double plastic bags at 220°C until the day of testing. Before testing, dual-energy X-ray absorptiometry was used to examine the variance in the bone mineral density (BMD) for each specimen. The thoracolumbar spine specimens were thawed to room temperature for 10 h and denuded of all residual musculature, with care taken to preserve the pertinent ligamentous structures and maintain segmental integrity.

Sha *et al.*[9] carried out a biomechanical study comparing a new combined rod-plate system with conventional dual-rod and plate systems. In general, the use of a dual-rod system allows for easier implantation and provides compensation for difficult anatomical and expositional conditions using adaptable and multidirectional screw positioning. Dual-rod systems also offer greater adjustability and control over screw placement, as well as increased load sharing and distraction and compression capabilities.

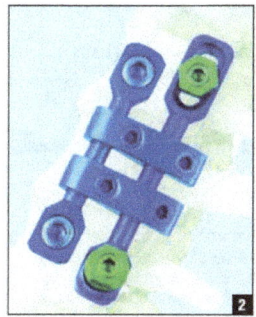

Fig. 6.17: Illustration of D-rod in the spine.

Fig. 6.18: An L1 corpectomy specimen with the Z-plate.

Source: http://www.healio.com/~/media/Journals/ORTHO/2013/2_February/10_3928_01477447_20130122_28/10_3928_01477447_20130122_28.pdf

Pelvis (Figs. 6.18 and 6.19)

Pelvic circumferential compression devices are designed to stabilise the pelvic ring and reduce the volume of the pelvis following trauma. It is uncertain whether pelvic circumferential compression devices can be safely applied for all types of pelvic fractures because the effects of the devices on the reduction of fracture fragments are unknown.

Knops *et al.*[10] conducted a study to compare the effects of circumferential compression devices on the dynamic realignment and final reduction of the pelvic fractures as a measure of the quality of reduction. Three circumferential compression devices were evaluated: the Pelvic Binder, the SAM Sling and the T-POD. In 16 cadavers, four fracture types were generated according to the Tile classification system (Fig. 6.20).

In the partially stable and unstable (Tile type B and C) pelvic fractures, all circumferential compression devices accomplished closure of the pelvic ring and consequently reduced the pelvic volume. No adverse fracture displacement (>5 mm) was observed in these fracture types. The required pulling force to attain complete reduction at the symphysis pubis varied substantially among the three different

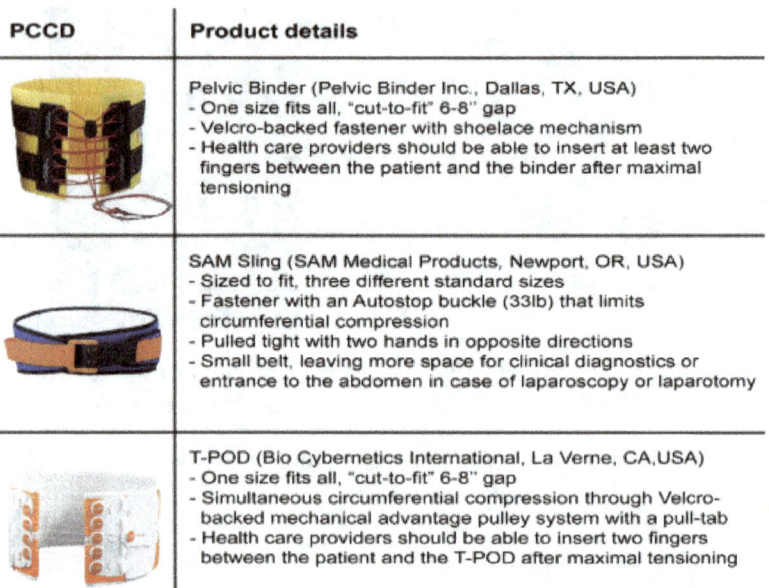

Fig. 6.19: The three commercially available pelvic circumferential compression devices evaluated in this study (Pelvic Binder, SAM Sling and T-POD) with the product details and manufacturers' guidelines for their application.

Source: http://www.medicalsca.com/files/jb__js_-_pelvic_devices_comparison_-_highlights.pdf

circumferential compression devices. The Pelvic Binder, SAM Sling and T-POD provided sufficient reduction in partially stable and unstable (Tile type B1 and C) pelvic fractures. No undesirable over-reduction was noted. The pulling force that was needed to attain complete reduction of the fracture parts varied significantly among the three devices, with the T-POD requiring the lowest pulling force for fracture reduction.

The results of this biomechanical cadaver study suggest that circumferential compression devices can provide early, noninvasive circumferential compression in partially stable and unstable pelvic fractures for advantageous realignment and reduction of these fractures without over-reduction. Clinical effectiveness of circumferential compression devices in patients with pelvic ring fractures remains to be determined.

Classification	Stability	Study definition
Tile A	Stable	A fracture in the os pubis was created 2 cm lateral from the symphysis pubis combined with a large fracture of the os ilium, ranging from the spina iliaca up to the tuber 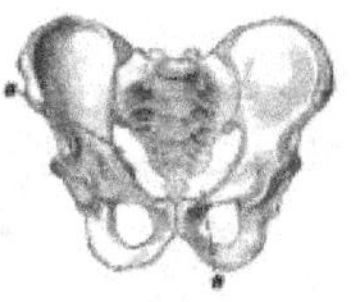
Tile B1 (50 mm) (100 mm)	Partially stable	A fracture in the os pubis was created through the symphysis pubis and displaced (50 or 100 mm) with a Finochietto rib spreader, causing unilateral rupture of the anterior ligaments of the SI-joint 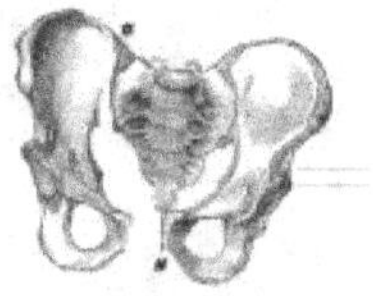
Tile C	Unstable	Complete pelvic ring instability was created through a fracture of the os pubis and a unilateral rupture of the SI-joint, including disruption of the soft tissue and rupture of the sacroiliac and sacrotuberous ligaments 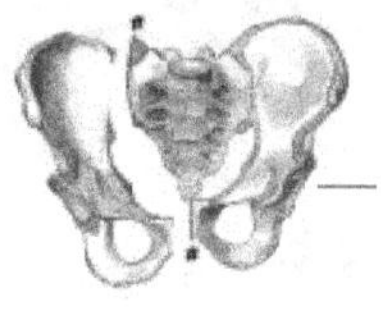

Fig. 6.20: The study definitions of the pelvic fractures according to the classification system of Tile et al.[11] A schematic representation of the study definition of the fractured pelves in this cadaveric study is shown in the right column.

Source: http://www.medicalsca.com/files/jb__js_-_pelvic_devices_comparison_-_highlights.pdf

Biomechanics Research

Type of Equipment Required

To perform biomechanical testing, the Department must have a Shimadzu Universal Testing Machine (Fig. 6.21) or an Instrom Machine. One should also collaborate with a bioengineer. Departments which lack such facilities and expertise should engage engineers from the Department of Mechanical Engineering. If not, such biomechanical studies cannot be performed.

NUS Orthopaedic Biomechanics Laboratory

The NUS Orthopaedic Biomechanics Laboratory is a service laboratory providing support to a wide range of research that requires

 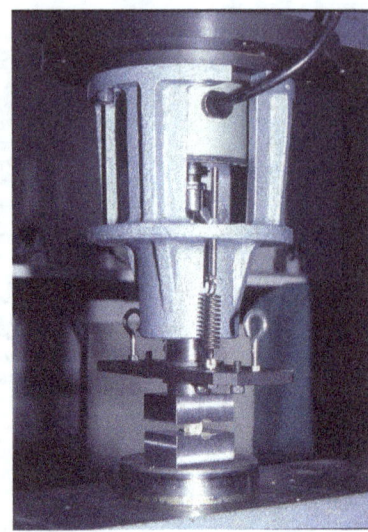

Fig. 6.21: Torsional testing using Shimadzu Universal Testing Machine Autograph DCS series with a torsion test device of 50 kg force metre.

validation of a model be it artificial, biological or artificial–biological construct.

The Orthopaedic Biomechanics Laboratory supports:

- A workshop for minor jigs modification
- Pressure sensing of area-contact under load (Tekscan and Fujifilm system)
- Laser system and video extensometer
- Motion analysis during biomechanical testing (Vicon motion capture system; Fig. 6.22)
- Biomechanical testing of small and large specimens (MTS 858 Mini Bionix biomechanical testing machines; Figs. 6.23a–c).

Collaboration with Mechanical Engineers

The biomechanics research engineer will help coordinate the various studies run in the biomechanical testing laboratory.

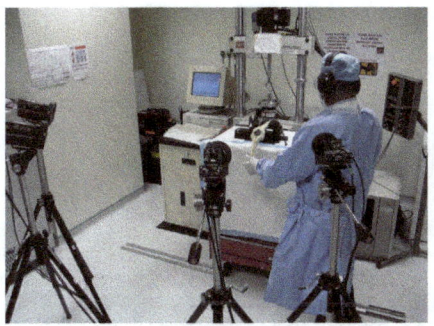 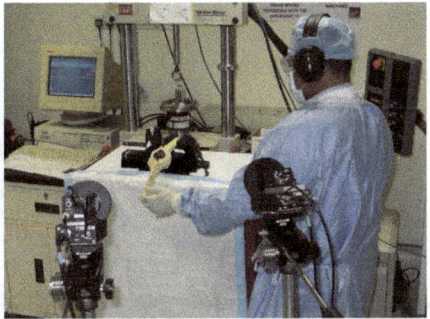

Fig. 6.22: Stimulating knee model range of motion with Vicon motion capture cameras used to track 3D motion.

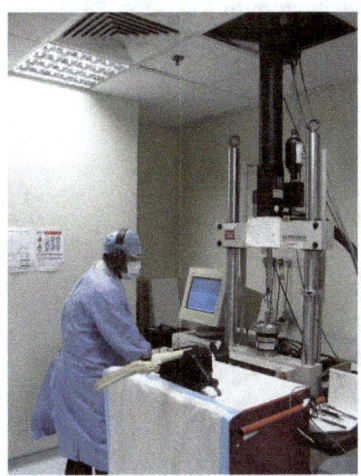

Fig. 6.23a: Operating the MTS 858 Mini Bionix testing machines with compressive, tensile and torsional testing functionalities.

The role of the engineer is to:

- Design, construct, calibrate, programme and test the device
- Conduct experiments utilising cadaveric specimens
- Collaborate with surgeons, scientists, statisticians and other research engineers

 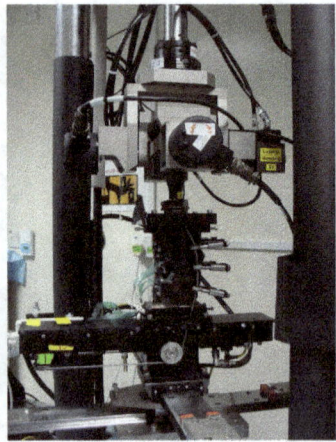

Fig. 6.23b: Securing specimen onto the MTS 858 Mini Bionix II with 6 degrees of freedom module to simulate the complex physiological motion of spine and other joint models.

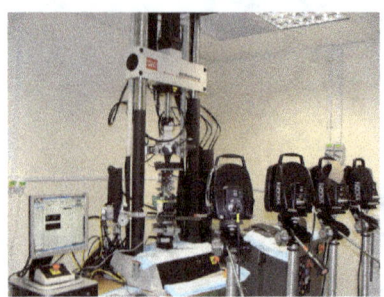 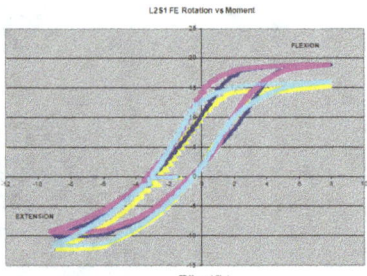

Fig. 6.23c: Spine testing with Tekscan sensor, Vicon cameras with reflective markers on specimen mounted on the MTS 858 Mini Bionix II with spine testing module and follower load system to simulate the complex physiological motion.

Source: http://medicine.nus.edu.sg/os/facilities/biomechanics/index.html

- Assist in day-to-day activities in the lab including test protocols and general lab duties
- Train and mentor students in the lab
- Support tissue retrieval and dissection for the cadaveric tissue used in the research studies

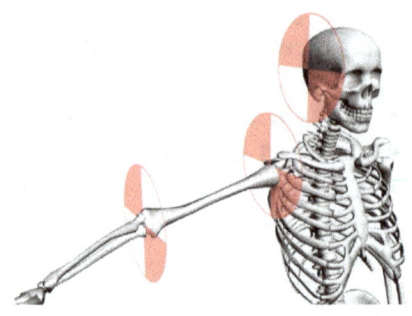

Type of Cadavers

Human

Cadavers that are used for medical and scientific research are donated or volunteered by their former owner or in some cases, by their family. No organisation, government or otherwise, has the ability to randomly confiscate human bodies for scientific research.[12]

Human cadavers are used by professionals to learn more about human anatomy (Fig. 6.24) and to gain information about diseases and disorders. Apart from medical research, human cadavers are also used in car-crash testing when dummies will not suffice, or in studies of decay for forensic purposes. Beating-heart cadavers are used for organ donations, especially ones that could not be preformed otherwise.

Human cadavers are used because the size, shape and exact location of organs vary from one individual to another. Organs are connected to other parts of the body in complex ways that textbook illustrations cannot effectively reproduce. Specimens of human cadavers can be classified into 'good' and 'less ideal'. A good specimen, in this context, means a young cadaver, one not overly obese or too evidently diseased.

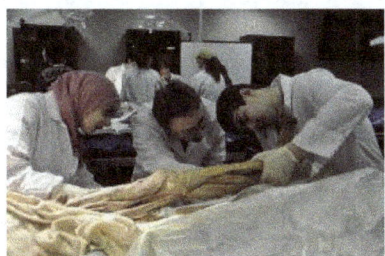 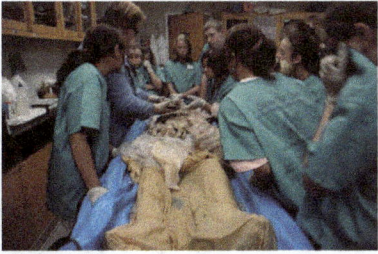

Fig. 6.24: Examination of cadavers.
Source: http://www.adn.com/2009/05/24/807234/the-dead-aid-the-living-in-uaa.html
Source: http://webs.purduecal.edu/nwi-ahec/past-events-2009/

Animals

Humans have a say in whether their bodies will be used for scientific research — animals do not. Animal specimens are brought in mostly from dissection farms, where they are raised specifically for the purpose that they will die. Animal cadavers, organs and tissue are usually obtained from sources where animals suffer harm or are killed, such as research facilities, animal breeders, farms, slaughterhouses, zoos, sporting events and some animal shelters (Fig. 6.25).[13]

Fig. 6.25: Animals confined in slaughterhouses and factory farms.
Source: http://www.green-blog.org/2010/07/22/the-cruel-life-inside-a-factory-farm/

However, cadavers, organs and tissue from species of animals which are less common than free-living wild, stray or companion animals, may be hard to source ethically (Fig. 6.26). In these situations, other sources of material, such as some of the above, may in certain circumstances provide an appropriate solution to the ethical challenge.

Fig. 6.26: Veterinary school cadaveric specimen of a horse head.
Source: http://svpow.com/category/stinkin-heads/

Areas of Research

Prosthesis

Basic, clinical, and applied research applications that examine the development and improvement of orthotic and prosthetic devices are of high priority. The field of biomechatronics is defined as the science of using mechanical devices with the muscular, skeletal or nervous system to assist or enhance motor control.[14] A prosthesis is an artificial device extension that replaces a missing body part.

Prostheses are typically used to:

- Replace parts lost by injury due to trauma;
- Replace parts missing from birth as a result of congenital diseases; or
- Supplement defective body parts.

Inside the body, artificial heart valves (Fig. 6.27) are in common use with artificial hearts and lungs which are under active technology development. Other medical devices and aids that can be considered prosthetics include artificial eyes, palatal obturator, gastric bands and dentures.

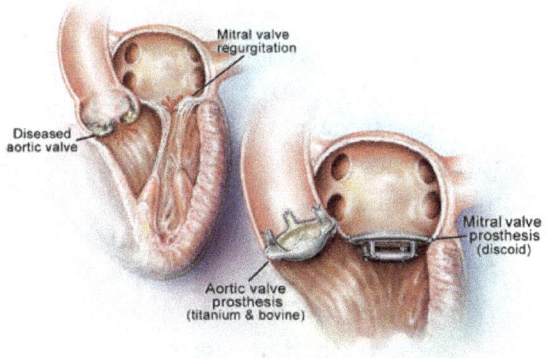

Fig. 6.27: Artificial heart valves.

Source: http://www.ptei.org/interior.php?pageID=83

The most common type of prosthesis would be artificial limbs, which can be further classified into four main types.

Transtibial prosthesis

A transtibial prosthesis is an artificial limb that replaces a leg missing below the knee. Transtibial amputees are usually able to regain normal movement more readily than an individual with a transfemoral amputation, largely due to retaining the knee, which allows for easier movement. Prosthetic devices commonly use silicone, urethane or elastomeric gels fit directed to the residual limb and hold the prosthetic device with or without pin locks. Elevated vacuum socket use is also on the rise and the intimate fit provides better blood flow to the residue limb for greater limb health for the amputee.

Transfemoral prosthesis

A transfemoral prosthesis is an artificial limb that replaces a leg missing above the knee. Transfemoral amputees can have a very difficult time regaining normal movement. In general, a transfemoral amputee must use approximately 80% more energy to walk than a person with two whole legs.[15] This is due to the complexities in movement associated with the knee. In newer and more improved designs, hydraulics, carbon fiber, mechanical linkages, motors, computer microprocessors and innovative combinations of these technologies are employed to give more control to the user.

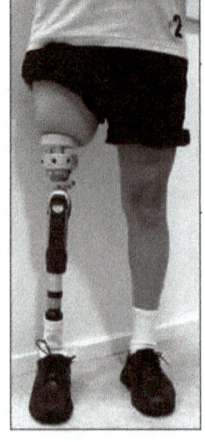

Transradial prosthesis

A transradial prosthesis is an artificial limb that replaces an arm missing below the elbow. Two main types of prosthetics are available: (1) cable-operated limbs, which work by attaching a harness and cable around the opposite shoulder of the damaged arm and (2) myoelectric arms, which work by sensing, via electrodes, when the muscles in the upper arm move, causing an artificial hand to open or close.

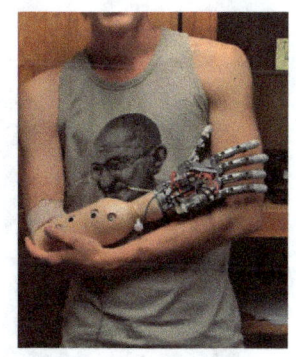

Transhumeral prosthesis

A transhumeral prosthesis is an artificial limb that replaces an arm missing above the elbow. Transhumeral amputees experience some of the same problems as transfemoral amputees, due to the similar complexities associated with the movement of the elbow.

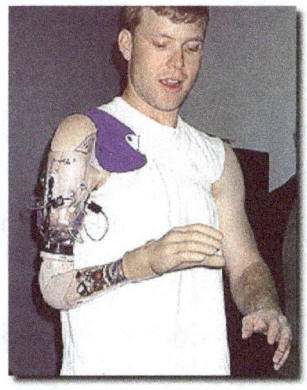

Plates/nails

An arthrodesis or fusion is an operation performed to fix a joint or joints in the foot and/or ankle. It may be used to treat a joint that is affected by severe arthritis or to correct deformity.[17]

Ankle fusion using screws

Your body is tricked into treating the joint as it would a broken bone. The joint surface is removed and screws or other metalwork are passed across the joint to maintain the position while the bone healing occurs (Figs. 6.28a and 6.28b). Bone then grows across the joint, fusing it solid. The aim of this operation is to turn a stiff painful joint

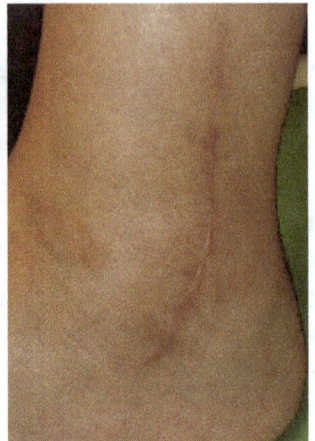
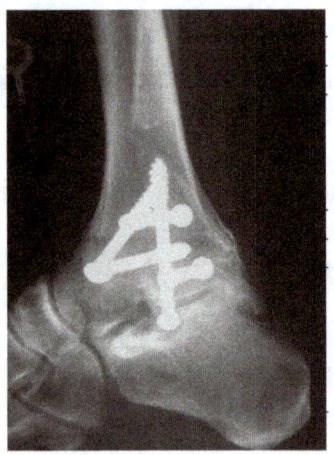

Fig. 6.28a: Post-operation scar.

Fig. 6.28b: X-ray showing ankle fusion using screws, wires and plates.

Source: http://www.georgelianmd.com/cms/Procedures/AnkleArthrodesis/tabid/66/Default.aspx

into a stiff painless joint. The operation is carried out only when all non-surgical measures have failed to control your pain.

Incisions are made over the front and on the inside of the ankle. The degenerate surfaces are cleared away and if necessary, reshaped to correct any deformity. The joint is placed into the correct position and fixed using screws. The ankle will then be protected by a plaster cast. Weight bearing will be limited until the bones knit together, after about three months post operation.

Intramedullary nail fusion

Occasionally, a procedure called a tibiotalocalcaneal fusion may be required. This involves fusing the shin bone (tibia) to the main bones in the back of the foot (talus and calcaneus).

The bones are usually fixed together using a large metal nail inserted into the middle of the shin bone. The nail is inserted through an incision in the bottom of the heel and screws are passed through the nail to prevent it from moving within the bone (Fig. 6.29).

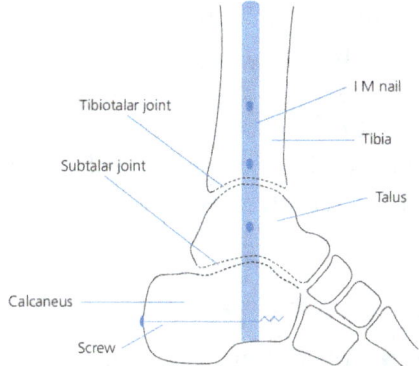

Fig. 6.29: Tibiotalocalcaneal fusion.
Source: http://www.georgelianmd.com/cms/Procedures/AnkleArthrodesis/tabid/66/Default.aspx

The main scar for this surgery will be on the outside of the ankle. Sometimes, a bone graft is required, particularly when a deformity needs to be corrected. This may either be taken from the bone that has already been removed during preparation of the joint surfaces, or from the pelvis or tibia.

Budget Costs

Different cadaveric parts have different charges. Even two of the same cadaveric organs might have different costs depending on factors such as age, sex, health status of the cadaver and whether specific requests have been made such as for removal of internal organs the torso, removal of brain from the head or shaving of hair from the head. Budgeting for transportation costs of the organs must also be taken into account.

Time Frame

The control cadavers for both the winter and summer trials were completely skeletal and started to undergo skeletal digenesis after the two-year period. Conversely, the experimental cadavers fell after a period of 1 to 2.5 years after placement with significant biomass remaining.

Understanding the variables that the hanging condition presents such as distance from the ground, positioning of the body in relation to the asphyxiation device, the type of noose used and how these affect the decomposition process, may allow us to better evaluate hangings in the outdoor context and hopefully lead to standards for estimating the postmortem interval in these contexts.

References

1. NUS Department of Orthopaedic Surgery. (2014). Orthopaedic Biomechanics Laboratory. Retrieved from http://medicine.nus.edu.sg/os/facilities/biomechanics/index.html
2. Poon, C. H. (2012). More cadavers needed, shortage may affect medical training. *The Straits Times*. Retrieved from http://medicine.nus.edu.sg/ant/body-donation/media/2012/ST_120428_More cadavers needed, shortage may affect medical training.pdf
3. Grossman, M. G., Tibone, J. E., McGarry, M. H., Schneider, D. J., Veneziani, S. & Lee, T. Q. (2005). A cadaveric model of the throwing shoulder: a possible etiology of superior labrum anterior-to-posterior lesions. *J. Bone Joint Surg.* **87**(4): 824–831.
4. McMahon, P. J., Chow, S., Sciaroni, L., Yang, B. Y. & Lee, T. Q. (2003). A novel cadaveric model for anterior-inferior shoulder dislocation using forcible apprehension positioning. *J. Rehabil. Res. Dev.* **40**(4). Retrieved from http://www.rehab.research.va.gov/jour/03/40/4/mcmahon.html
5. Telleria, J. J. M., Lindsey, D. P., Giori, N. J. & Safran, M. R. (2011). An anatomic arthroscopic description of the hip capsular ligaments for the hip arthroscopist. *Elsevier* **27**(5): 628–636.
6. Tan, J. L., Chang, P. C. C., Mitra, A. K. & Tay, B. K. (1998). Anthropometry of anterior cruciate ligament in Singaporean Chinese. *Ann. Acad. Med.* **27**(6): 776–779.
7. Nester, C. J. (2009). Lessons from dynamic cadaver and invasive bone pin studies: do we know how the foot really moves during gait? *J. Foot Ankle Res.* **2**: 18.

8. MacAvoy, M. & Green, D. P. (2007). Critical reappraisal of Medical Research Council muscle testing for elbow flexion. *J. Hand Surg. Am.* **32**(2): 149–153.
9. Sha, M, Ding, Z.-Q., Ting H. S., Kang L.-Q., Zhai, W.-L. & Liu H. (2013). Biomechanical study comparing a new combined rod-plate system with conventional dual-rod and plate systems. *Orthopedics* **36**(2): e235–e240.
10. Knops, S. P. (2010). Comparison of three different pelvic circumferential compression devices: A biomechanical cadaver study. *J. Bone Joint Surg. Am.* **93**(3): 230–240.
11. Tile, M., Helfet, D. L. & Kellam, J. F. (2003). *Fractures of the Pelvis and Acetabulum* (3rd ed.). Philadelphia: Lippincott Williams & Wilkins.
12. Donaghey, J. (2007). Human cadaver dissection for medical and scientific research *vs.* animal specimen dissection in grade school classrooms. Retrieved from http://www.scribd.com/doc/7500305/Human-Cadaver-Dissection-for-Medical-and-Scientific-Research-vs-Animal-Specimen-Dissections-in-Grade-School-Classrooms
13. Jules, N. (2011). InterNICHE policy on the use of animals and alternatives in education and training. Retrieved from http://www.interniche.org/en/system/files/public/Policy/InterNICHE_Policy_v3.0_-_Nov_2011.doc.
14. Ottobock. (1998). Prosthetic arms and legs fitting users in India. Retrieved from http://www.ottobock.in/glossary.asp
15. Vanderwerker, Jr., E. E. (1976). A brief review of the history of amputations and prostheses. *Inter-Clin. Inf. Bull.* **15**(5): 15–16.
16. Alligan, K. (2011). A patient's guide to ankle and hindfoot arthrodesis or ankle and hindfoot fusion. Retrieved from http://www.rnoh.nhs.uk/sites/default/files/patient/11-177_rnoh_amendments_10251_pg_ankle_and_hindfoot_web_release.pdf

Section III
Ethics and Statistics

Chapter 7

Ethics for Research

Joy Le Yi Wong & Aziz Nather

> *"Ethics is knowing the difference between what you have a right to do and what is right to do."*
>
> Potter Stewart (1915–1985)[1]

What is Research Ethics?

Ethics is a cornerstone for conducting meaningful and fulfilling research. There has been much thought and debate about the boundaries that researchers are allowed to explore. Nevertheless, it is imperative that one understands a few basic concepts to guide one's research endeavour.

Research ethics is a set of ethical principles that guide and govern the design and implementation of research. Ethics are norms of conduct that distinguish between what is morally sound and what is unacceptable.[2]

Why is ethics important? Ethics have the potential to shake up the very foundations of research. Research is built upon **trust**. Researchers and society must trust that the results of research are accurate and without bias. For this trust to endure, the scientific community must exemplify the values that the ethics community uphold.

Ethics also ensure that researchers can be held **accountable** to the public. Cases of data fabrication and human subject abuse can be avoided if researchers adhere to the set of ethical guidelines laid out by scientific communities.[3]

Finally, ethics also maintain the **sanctity and protection** of human and animal subjects. These guidelines ensure that their welfare is not compromised and that no subjects are harmed in the process of experimentation.[4]

 This article discusses the following areas of research:

1. Human experimentation
2. Animal experimentation
3. Cadaver experimentation
4. Boundaries in research
5. Academic scandals

HUMAN EXPERIMENTATION

Background of ethics

1. Official ethical guidelines include:

i) The Hippocratic Oath
ii) Declaration of Geneva
iii) Nuremberg Code
iv) Declaration of Helsinki
v) The Belmont Report

i) The Hippocratic Oath

> **Hippocratic Oath**
>
> I do solemnly swear by that which I hold most sacred:
>
> That I will be loyal to the profession of medicine and just and generous to its members;
>
> That I will lead my life and practice my art in uprightness and honor;
>
> That into whatsoever house I shall enter, it shall be for the good of the sick to the utmost of my power, I holding myself aloof from wrong, from corruption, and from the temptation of others to vice;
>
> That I will exercise my art solely for the cure of my patients, and will give no drug, perform no operation for a criminal purpose, even if solicited, for less suggest.
>
> That whatsoever I shall see or hear of the lives of men which is not fitting to be spoken, I will keep inviolably secret.
>
> These things I do promise, and in proportion as I am faithful to this my oath may happiness and good repute be ever mine,— the opposite if I shall be forsworn.

The Hippocratic Oath (Ορκος) is perhaps the most famously known of the Greek medical texts. Its principles are held sacred by physicians and healthcare professionals to this day.[5]

The Hippocratic Oath can be summarised as:

— Unity among the healthcare teachers and professionals
— To do no harm
— To not assist suicide or abortion
— To not be ashamed to say "I know not"
— To respect the privacy of patients
— To not play God[6]

ii) Declaration of Geneva

The Declaration of Geneva was intended as a revision of the Hippocratic Oath to ensure that it retains its relevance in the modern-day context. The Declaration was adopted by the General Assembly of the World Medical Association (WMA) in 1948. It was written after World War II, declaring the ethical goals of medicine.[7]

The Declaration of Geneva as published by the WMA reads:[8]

> At the time of being admitted as a member of the medical profession:
> - I solemnly pledge to consecrate my life to the service of humanity;
> - I will give to my teachers the respect and gratitude that is their due;
> - I will practice my profession with conscience and dignity;
> - The health of my patient will be my first consideration;
> - I will respect the secrets that are confided in me, even after the patient has died;
> - I will maintain by all the means in my power, the honour and the noble traditions of the medical profession;
> - My colleagues will be my sisters and brothers;
> - I will not permit considerations of age, disease or disability ... or any other factor to intervene between my duty and my patient;
> - I will maintain the utmost respect for human life;
> - I will not use my medical knowledge to violate human rights and civil liberties, even under threat;
> - I make these promises solemnly, freely and upon my honour.

iii) Nuremberg Code

The Nuremberg Code is a set of research ethics principles for human experimentation as a result of the Nuremberg trials after World War II.

> Historical Context: The code was drafted in response to the atrocities committed by the German doctors during the Holocaust. A German doctor, Joseph Mengele, also known as the "Angel of Death", was infamous for his cruel and inhumane experiments involving Auschwitz twins. He conducted various surgeries without anaesthesia such as organ removal, castration and amputations.[9]

The **10** key points of the Nuremberg Code are:

1. The importance of **voluntary consent of the human subject**
2. Experiments should **yield fruitful results** and are not random and frivolous
3. Experiments should be designed based on **prior research** and animal experimentation
4. Unnecessary **physical and mental suffering** is to be avoided
5. No experiment should be conducted if **death** or disabling injury is expected
6. The **degree of risk** should never exceed the significance of the experiment
7. **Proper preparations** should be made to protect the subject from harm
8. Experiments can only be conducted by **scientifically qualified persons**
9. Human subjects have the **freedom to withdraw** from the experiment
10. Scientists in charge must be ready to **terminate the experiment** if it is likely to results in any grievous harm to subjects.[10]

iv) Declaration of Helsinki

The Declaration of Helsinki is a set of ethical principals regarding human experimentation developed by the WMA. It is commonly regarded as the cornerstone document of human experimentation. The rules mainly revolve around the protection of the individual and his rights.[11]

v) The Belmont Report

The Belmont Report, titled *The Belmont Report: Ethical Principles and Guidelines for the Protection of Human Subjects of Research, Report of the National Commission for the Protection of Human Subjects of Biomedical and Behavioral Research*, is arguably one of the most widely regarded code of ethics. It was created by the National Commission for the Protection of Human Subjects of Biomedical and Behavioral Research and drafted in Belmont Conference Center in Elkridge, Maryland.

The Report highlights three main guidelines for human research: respect for subjects, beneficence and justice. The first emphasises the need for autonomy of human research subjects. Groups without sufficient autonomy such as children must be given special protections and considerations. The second, beneficence, refers to the need for the contributions of the study to outweigh any harm sustained by the participants. Lastly, justice refers to the treatment of vulnerable populations, and the equal distribution of the benefits and findings of the research.[12]

2. Keeping Secrets!

Confidentiality of Patient Information

As a researcher, you have access to sensitive personal information about the subject. It is your duty to keep this information confidential unless consent is given by the patient.

> *"Respect the principle of medical confidentiality and not disclose without a patient's consent."*
>
> — The Singapore Medical Council (SMC) in *Ethical Code and Ethical Guidelines* (2002)

Vulnerable Groups

> Confidentiality is at the centre of maintaining trust.

The confidentiality of people, especially those living with "stigmatised disease" such as HIV/AIDS, mental illnesses and substance abuse should be taken care of. It is important that the researcher recognises their concerns and ensures that their privacy is protected.[13]

Benefits of Maintaining Confidentiality

- Establishes trust between the researcher and subjects
- Reduces unwanted anxiety on the part of the individual
- Preserves the individual's dignity
- Ensures that the participant feels respected
- Gives the participant control and autonomy

Participation in research is voluntary, thus a researcher's obligation to protect patient confidentiality is of paramount importance. People will not volunteer for research unless they are certain that the information they disclose will not be revealed to the public without their consent.[14]

Confidentiality and the Greater Good

While subject confidentiality is important, there might be situations where the researcher can reveal information for the patient's own good.

For example, if a researcher is aware of abuse in a care home, it would be right to inform the relevant authorities. Researchers are advised to act in accordance to one's own moral judgement.[15]

The Ethics and Confidentiality Committee (ECC) was set up under Section 251 of the National Health Service (NHS) Act 2006 to advise on ethical issues regarding medical information.

> *"Allow the common law duty of confidentiality to be set aside in specific circumstances where anonymised information is not sufficient and where patient consent is not practicable."*[16]
>
> — Section 251 of NHS Act (2006)

3. Informed Consent

> *"Every human being of adult years and sound mind has a right to determine what shall be done with his own body; and a surgeon who performs an operation without his patient's consent, commits an assault, for which he is liable in damages."*[17]
>
> — Justice Benjamin Cordozo (1914)

Informed consent is ensuring that all potential subjects are fully aware of all the important information about the experiment, including the risks, benefits and procedure. Information should be conveyed in a clear and transparent manner, minimising ambiguity. Subjects must have a complete understanding of the experiment that they are volunteering for.

It is the responsibility of the clinician to enhance the patient's capacity by minimising language barriers. Impaired senses such as poor hearing and eyesight must be accommodated for. Ample time should also be given to the patient to make his or her decisions. Information provided should be in clear and simple language, without jargon or overly technical terms.[18]

Subjects are also allowed to withdraw from the experiment at any point of time and for any reason. This is a basic and legal ethical standard by which all research must adhere to.

The subject must know that participation is entirely voluntary.

Researchers have the duty to make sure that participants are well aware of the following before they consent to take part:

- What will their participation entail?
- What are the possible risks?
- What are the possible benefits?

- What are the possible alternatives to participating?
- What are the rights to confidentiality and privacy?
- What compensation is there for possible injuries incurred during the study?[16]

> *"It is essential that the information provided is understood by the potential participant and empowers that person to make a voluntary decision about whether or not to participate in the study."*[17]
>
> — Family Health International (2009)

Table 7.1 shows a summary of the laws in Singapore with regard to consent of human patients in clinical contexts.[18]

Table 7.1: Summary of the law in Singapore with regard to age and consent

Under the Penal Code (Chapter 224, 2008)
1. Consent by parents and guardians needed for children below 12 years (Section 89)
2. Valid consent by persons above the age of 18 (Section 87)
3. Acts done in good faith for the benefit of a person without consent (Section 92)

Under the Civil Law (Amendment) Act (Chapter 43, 2009)
1. Confers contractual capacity to persons aged 18 and above

Under the Children and Young Persons Act (Chapter 38)
1. Juvenile; a person who is seven years or above, and below 16 years
2. Child: a person below the age of 14 years
3. Young person: a person 14 years or above, and below 16 years

Summary
1. Ages below 14: needs consent from a person of parental responsibility
2. Ages 14 to 16: Gillick competence may apply (Gillick v West Norfolk and Wisbech Area Health Authority [1986] AC 112)
3. Ages 16 to 18: persumed to be able to consent for medical treatment unless proven otherwise
4. Ages 18 and above: may consent for necessary medical treatment
3. Age 21: the age of the majority

Source: http://sma.org.sg/UploadedImg/files/Publications%20-%20SMA%20News/4508/CMEP.pdf

ANIMAL EXPERIMENTATION

Each year, more than 100 million animals are used for experimentation. These animals include mice, rats, frogs, dogs, cats, rabbits, hamsters, monkeys and birds. The animals are used for various reasons, ranging from biology studies to quality control of products, as shown in the figure below.

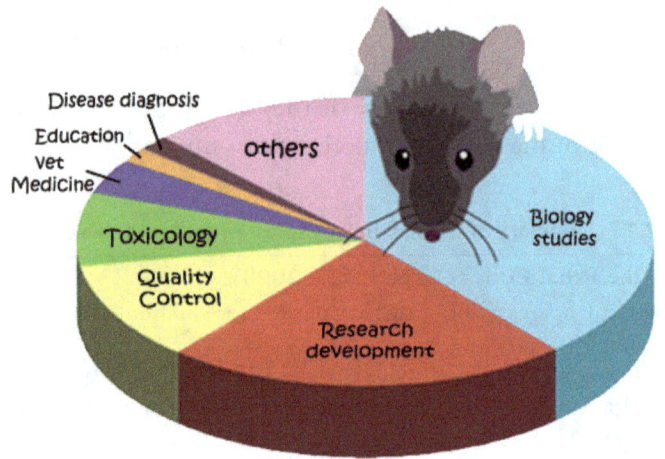

Source: http://www.nature.com/nm/journal/v16/n11/full/nm1110-1172b.html

"Animal research has played a vital role in virtually every major medical advance of the last century — for both human and veterinary health. From antibiotics to blood transfusions, from dialysis to organ transplantation, from vaccinations to chemotherapy, bypass surgery and joint replacement, practically every present-day protocol for the prevention, treatment, cure and control of disease, pain and suffering is based on knowledge attained through research with lab animals."[19]

— Foundation for Biomedical Research (2013)

While animals have indeed helped mankind on the march towards progress, it would be prudent to first consider the flaws of animal experimentation and lay out the necessary guidelines regarding animal research.

History of Animal Experimentation

Early History

Animal dissection has been taking place since circa 500 BC, scientists such as Aristotle, Herophilus and Erasistratus performed such experiments to uncover the anatomy and functions of living organisms. Laws against mutilation of the human body in ancient Rome and Alexandria resulted in a reliance on animal subjects.

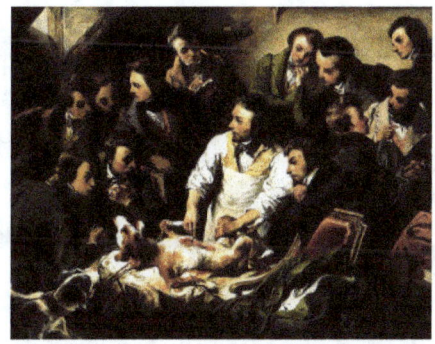

Source: http://animal-testing.procon.org/#background

The French philosopher René Descartes occasionally experimented on live animals. He believed that animals were "automata", unable to experience pain as humans do. He acknowledged that animals could feel but since they were unable to think, they could not consciously experience emotions.

> *"It seems reasonable since art copies nature, and men can make various automata which move without thought, that nature should produce its own automata much more splendid than the artificial ones. These natural automata are the animals."*[20]
>
> — René Descartes in a letter to Henry More (February 5, 1649)

Queen Victoria was an opponent of animal testing in England. A letter by her private secretary in 1875 read:

"The Queen has been dreadfully shocked at the details of some of these [animal research] practices, and is most anxious to put a stop to them."

Source: http://www.goldposters.com/item-1867311/queen-victoria-cartoon.html

Thus, in 1876, the Great Britain's Cruelty to Animals Act was established.

In recent years, animal experimentation has come under heavy criticism from animal rights groups. Many countries have passed laws to ban or limit animal experimentation. Debates about the ethics of animal testing have been raging on since the 17th century.[21]

The Pros and Cons of Animal Testing
PROS

1. Aids in research for new drugs and treatments

Many medical breakthroughs have been made possible through animal testing, including treatments for cancer, HIV, antibiotics and vaccines. The California Biomedical Research Association stated that almost every medical breakthrough in the last 100 years involved research using animals.

Cancer

Animal research helped to develop *Herceptin* and *Tamoxifen*, drugs that have saved thousands of patients suffering from breast cancer.

The discovery of a protein that caused leukemia in mice led to the development of *Gleevec*, the first molecularly targeted drug against cancer. It was approved by the U.S. Food and Drug Administration (FDA) in May 2001 for the treatment of chronic myeloid leukaemia (CML) and a previously incurable form of stomach cancer known as gastrointestinal stromal tumour (GIST).

Diabetes

Millions of lives have been saved because of the discovery of insulin in dogs. In 1889, German scientists Oskar MInkowski and Joseph Von Mering removed a pancreas from a dog for research on its digestive function. After the pancreas was removed, flies were observed to swarm around the dog's urine. The scientists discovered that the pancreas secreted hormones that regulated blood sugar, providing a model for the study of diabetes.

Several new treatments, such as quick-acting and long-acting insulin have also been developed using animal models.[22]

2. Ensures safety of drugs

In 1937, a US pharmaceutical company created a drug called 'Elixir Sulfanilamide', which contained diethylene glycol (DEG) as a solvent. No animal testing was done and the company's chief pharmacist was not aware that DEG was poisonous to humans. The company sold the drug and resulted in a mass poisoning of more than a hundred people. The incident led to a public outcry which resulted in the passing of the 1938 Federal Food, Drug and Cosmetic Act, which requires safety testing of drugs on animals before they could be marketed. The above-mentioned incident illustrates the possible harm to humans from the use of drugs that have not been pre-tested on animals. Animal testing in biomedical research has led to tremendous advances in the treatment of various diseases.[23]

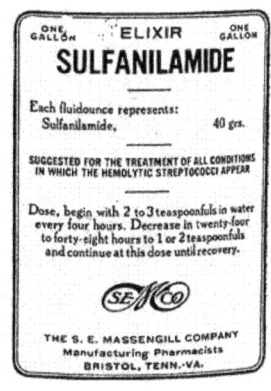

3. Animals are genetically similar to humans

Chimpanzees and humans share almost 99% of their DNA, and mice are 98% genetically similar to humans. Thus, animal experimentation provides a viable model to test novel drugs and treatments instead of conducting human trials.

> "Medical research involving human subjects must … be based on a thorough knowledge of the scientific literature … and, as appropriate, **animal experimentation**."
>
> — Declaration of Helsinki (2013)

4. Animals do not have equal rights as humans

Do animals have the same rights as humans?

Some argue that since animals do not have the moral judgement that humans do, they do not command the same rights that humans have. Thus, for the greater good of mankind, some animals need to be used during experiments to allow for the advancement of science and research.[24]

Billboard by the Foundation for Biomedical Research.

CONS

1. Animal testing is inhumane

According to Humane Society International, animals are often subjected to force feeding, infections and kept at extreme conditions to study the infection and healing processes. The U.S. Department of Agriculture (USDA) reported in 2010 that 97,123 animals suffer pain without anaesthesia, including 1,395 primates, 5,996 rabbits, 33,652 guinea pigs and 48,015 hamsters.

2. Animals have rights too

> *"The greatness of a nation and its moral progress can be judged by the way its animals are treated."*
>
> — Mahatma Gandhi

Some argue that animal and human rights boil down to one fundamental principle: the right to be treated with respect as an individual with value.

Contrary to what some believe, animals do show altruistic behaviour. Jane Goodall, an expert on chimpanzees, reported that animals sometimes show genuinely altruistic behaviour. Thus, it is unethical to assume that since animals do not seem to show the same complex cognitive abilities as humans then they are of a lower class than humans. Such discrimination between humans and animals is called "speciesism", as coined by Peter Singer. He defined it as:

> *"A prejudice or bias in favour of the interests of members of one's own species and against those of members of other species."*[25]

— Peter Singer, Animal Liberation (1975)

3. Animal tests are unreliable

A 2013 study in the *Archives of Toxicology* stated that: "The low predictivity of animal experiments in research areas allowing direct comparisons of mouse versus human data puts strong doubt on the usefulness of animal data as key technology to predict human safety." A scientific study reported that 94% that enter clinical trials following animal experimentation fail to be approved. Of the remaining that are approved, most are withdrawn due to lethal side effects that were not detected during animal studies.[20]

A few basic principles of animal experimentation

i. **Choice of organisms**
   Preferential use of less complex organisms should be practiced whenever possible. For example: bacteria and plants would be preferred over mammals.

ii. **Reduce animal use as far as possible**
   Extensive literature searches ensure that experiments are not unnecessarily replicated. Animal models should only be used to obtain information not already available in the scientific community.

iii. **Best possible treatment of animals used in a study**
High-quality and disease-free experiment environments should be provided during animal experimentation. Anaesthesia should be used when appropriate, to minimise pain for animals.[26]

iv. **3Rs Principle of Replacement, Reduction and Refinement**
The 3Rs was first introduced by William Russell and Rex Burch in *The Principles of Humane Experimental Technique* in 1959. They have been internationally accepted as the basis of the care and treatment of animals. These are to:

Replace the need for animal experimentation by alternative means
Reduce the numbers of animals used to a minimum
Refine the procedures so as to minimise the harm inflicted on animals

National Advisory Committee for Laboratory Animal Research (NACLAR)

Guidelines on the Care and Use of Animals for Scientific Purposes

NACLAR was set up in 2003 to set out the framework for the guidelines for the responsible use of animals for scientific purposes in Singapore.

➢ Use animals bred in captivity over those taken from their natural habitats.
➢ Projects must be designed to minimise pain to animals.

> Procedures that result in pain for animals should be performed with anaesthesia.
> If severe pain cannot be alleviated, the animal must be killed humanely.

NACLAR places strong emphasis on the 3Rs principle of Replacement, Reduction and Refinement, as mentioned above, which will be further elaborated upon.

Replace: Alternative methods, such as computer stimulation, *in vitro* biological systems and 3D models must be considered before embarking on a project involving animals.

Reduction: The number of animals used must be the minimum number required to collect accurate results.

Refinement: Animals of the appropriate species must be chosen, taking to account their biological characteristics and genetic predispositions.[26]

The Institutional Animal Care and Use Committee (IACUC)

Every institution that uses animals for research must have an Institutional Animal Care and Use Committee (IACUC). Each local IACUC review research protocols and evaluates the treatment of animals during research. Before scientists embark on an animal research project, they must first obtain IACUC approval. Applications submitted by the 7th day of the month will be reviewed in the same month.[27]

PROCESS FLOWCHART FOR REVIEW OF IACUC APPLICATIONS

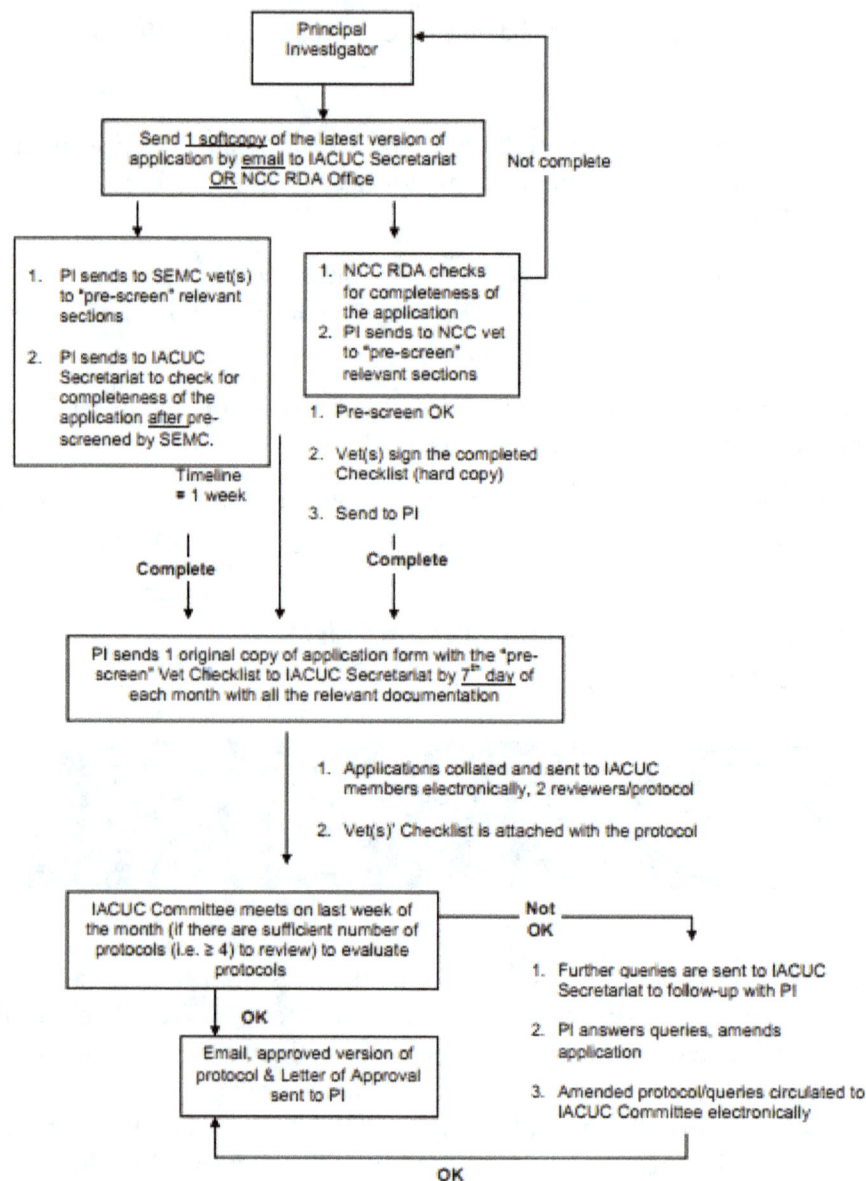

Source: http://research.singhealth.com.sg/Pages/InstitutionalAnimalCareandUseCommittee.aspx

CADAVERIC RESEARCH

Respect for the human body

The main ethical concern of using cadavers in research is respect for human life. Working with human material requires utmost respect and sensitivity. The following guidelines can help in understanding the responsibilities of using human material.

- No body part or tissue should be removed from the lab
- No photographs should be taken in the lab
- No friends or family should be allowed into the laboratory

Guidelines for handling the cadaver

- The cadaver should be cared for at all times. The cadaver has to be kept moist with towels dampened with embalming fluid
- Only unwrap the area that you are working on
- Periodically spray the area you are studying with embalming fluid
- When research is done, replace the towels and cover the cadaver with a plastic sheet
- Cadavers are embalmed with fluid containing glycerine, ethyl alcohol and phenol. Wear gloves at all times to avoid contact
- All cadaveric tissues must be well kept in containers[28]

WHERE DO WE DRAW THE LINE?

Genetic Engineering

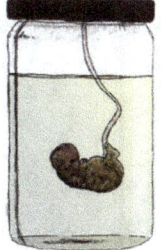
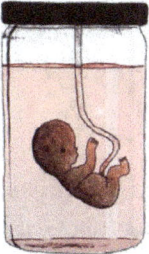

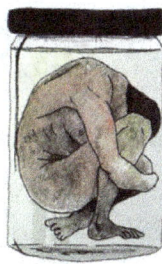

> *"The genetic code is 3.6 billion years old. It's time for a rewrite."*[29]
>
> — Tom Knight
> Professor at MIT's Artificial
> Intelligence Lab

The pantheon of living organisms on Earth is about to get some newcomers. Scientists in the last couple of years have been trying to engineer novel forms of life by tweaking the genetic code. Great breakthroughs have been made over the years in the field of genetic engineering and Nature's monopoly on creation has been constantly defied.

Yet, the issue of genetic engineering remains a largely controversial topic. The moral conundrum of where to draw the line in research remains unresolved.

At the end of the day, it would be wise to ponder over what our motives are in genetic engineering, and whether there are better biomedical ventures we should be exploring with our research funds. Before that, let us first examine some of the ethical implications of genetically modified organisms.[30]

i) Playing God

The issue of the acceptability of the genetic manipulation of animals and plants has always been a contentious one. Do we have the right to "play God", and tamper with nature by mixing genes among species? Is this a violation of the intrinsic rights of an organism? Many detractors believe that Nature created the best of possible worlds. Charles Darwin believed that the myriad designs of creation are perfectly shaped to be just the way they are. Yet, on the other end of the spectrum, there are some that are constantly pushing the boundaries of scientific progress and experimentation. As James Watson, co-discoverer of the structure of DNA says: *"If we don't play God, who will?"*

ii) Religious concerns

Pope Benedict XVI has condemned scientists who *"modify the very grammar of life as planned and willed by God"*. He once said during a Good Friday address that genetic engineering is a kind of "*anti-Genesis*" activity. Pope Benedict also prayed against "*insane, risky and dangerous*" ventures "*to take God's place without being God*".[27]

iii) Welfare of organisms

There is little concern whether genetically modified animals are biologically "designed" to withstand additional burden of increased milk or meat production. For instance, bovine somatotropin (BST), a protein hormone in dairy cattle allows cows to produce more milk, but at the same time increases the risk of mastitis — inflammation of the mammary gland.

iv) Exploitation of the poor

Are richer countries benefitting at the expense of poorer ones? Worldwide food production may be dominated by a small number of large companies with the capital and knowledge to monopolise the market, resulting in over-reliance of the poor countries on the rich ones. Some companies have been pushing for "Terminator seeds", plants that are genetically designed to be unable to germinate a second time. This forces farmers to buy a new batch of seeds each year, many who are unable to afford the high costs of doing so. In India, there are fears for the livelihood of 400 million farmers and their agricultural produce.[31]

v) Are genetically modified organisms (GMOs) safe?

Whilst it is tempting to declare that genetically modified food, with its higher yields and anti-pest properties shall be the answer to the global food shortage, it remains a premature assertion at best. We still do not know the potential health risks of genetically modified foods, and no one can anticipate the potential consequences of releasing GMOs into the environment.

Our notion of what is right or wrong with our presently available scientific knowledge could easily change as society progresses. We should instead emphasise the need for constant debates to direct policies and frameworks within which genetic engineering can progress. Science should raise ethical issues and ethical debates should influence science, creating a healthy balance of research and ethical checks to keep scientific progress fruitful and fulfilling for all of mankind.[32]

Gene Therapy

> *"Should experiments in gene therapy be stopped when they have already helped some victims of rare diseases? Should the manufacture of genetically engineered hormones be stopped — even if it means depriving those who desperately need them?"*[33]
>
> — Linda Tagliaferro
> In *Genetic Engineering: Progress or Peril?*

Gene therapy has advanced tremendously in the recent years, providing unprecedented treatments to once incurable diseases. However, it is imperative to weigh the ethical implications of gene therapy before the "horse" bolts from the stable and it is too late to shut the doors.

i) Therapy vs. enhancement

There is a fine line between enhancement and treatment, the question is, where do we draw the line? How do we define what is "inferior" and what is "normal"? Once DNA is manipulated, what would be the repercussions on the new standards of talent and beauty?

ii) Eugenics

Critics argue that gene therapy will inevitably lead to the practice of eugenics, a purposeful effort to control the genetic makeup of human populations. It might even lead to selective breeding, to emphasise certain attractive traits in humans. The idea of forcing intelligent people to mate to produce offspring with high IQ seems outright unacceptable, but on further consideration, this might be something that people do voluntarily.

iii) Germline therapy

The Clothier committee on the Ethics of Gene Therapy reported, "We share the view of others that there is at present insufficient knowledge to evaluate risks to future generations." They advised that germline therapy "should not yet be attempted" as the risks to future generations are too huge.[34]

Furthermore, elimination of unwanted genes from the gene pool can possibly be detrimental to the future generation by reducing genetic variation. Another possible consideration is that genes that are harmful now might be advantageous in the future under different conditions.[35]

Embryonic Stem Cells

The 14-day rule of irreversible individuality of the embryo has of late been rendered problematic. There has been much debate about the "individuality" of the embryo and whether embryonic stem cell research is permissible. Two irreconcilable schools of thought championed by opposing camps of scientists inadvertently create a sense of disorientation in the scientific committee. Embryonic stem cell research poses a moral dilemma, forcing us to choose between two moral principles

→ To alleviate suffering
→ To respect the value of human life

For embryonic stem cell research, it is inevitable that the embryo is destroyed in the process. While it means destroying a human life, research also has the potential to discover new medical treatments and cures. So which principle should have the upper hand? To answer that question, we have to first consider if the embryo has the status of a person.

What moral status does the embryo have?

The moral status of an embryo is a highly complex issue. The main viewpoints are outlined below.[36]

Arguments for this view	*Arguments against this view*
1. The embryo is fully human upon fertilisation	
A fertilised egg will eventually develop into a full grown human and it is arbitrary to pinpoint when the baby is considered as "human".	A fertilised egg that has not yet been implanted in the uterus has not developed the psychological or physical properties of a human being. It is unable to feel, think or resemble a human.
Destroying the embryo during research is equivalent to sacrificing a person for the sake of research.	The embryo cannot develop into a child without being embedded in a uterus. Something that has the potential to become a person should not be treated as if it were actually a person.
2. The embryo gains individuality at 14 days after fertilisation	
After 14 days, the embryo can no longer split to form twins. If we say that the embryo is a human, then we are also saying that one of us can become two, which is impossible. At around 14 days, the central nervous system starts to develop, giving the embryo senses and feelings, making it more human-like.	Simply because the embryo was one before it twinned does it deny it the right of being human. The argument that individual life is not present is tenacious.
3. The embryo has increasing status as it develops	
There are certain stages of development that could be assigned moral status: 1. Implantation of the embryo into the uterus wall 2. Appearance of the primitive streak (approximately 14 days after fertilisation) the development of the central nervous system	The intrinsic value of a human lies in the life itself. If we judge the value of a human life by its age, then we are making unfair judgements of who is human.

3. Premature baby with physical human characteristics 4. Birth We tend to feel differently about a lost life depending on the stage where it is lost. More than half of the fertilised eggs are lost due to natural causes, if they are not areas of concern, then the usage of such eggs in research should not be a problem.	If we say that the formation of the nervous system marks a human being, can we say the same for a person that has lost control over his nervous system to be less human?
4. Embryos have no moral status	
Fertilised eggs are merely organic material with status equal to other body parts. It is not considered a living entity until it is able to survive independently on its own.	By destroying an embryo, we are removing all potential of it developing into a human being, which is tantamount to destroying a human life.

Ethics gone wrong

... *Academic scandals*

What is fabrication of data?

Research misconduct is defined by the **Royal College of Physicians of Edinburgh,** *"as any behaviour by a researcher, whether intentional or not, that fails to scrupulously respect high scientific and ethical standards. Various types of research misconduct include fabrication or falsification of data, plagiarism, problematic data presentation or analysis, failure to obtain ethical approval by the Research Ethics Committee or to obtain the subject's informed consent, inappropriate claims of authorship, duplicate publication, and undisclosed conflict of interest."*[37]

Fabrication of data[38] includes:

i) False citations, acknowledgements or incorrect documentation of sources
ii) Falsified, invented or fictitious data

iii) Concealment or omission of information
iv) Unauthorised submission of work prepared by another

Types of data fabrication

i) Duplicate submission

The submission of an article by two journals that overlap significantly or are identical are called duplicate publications. When an article is republished as part of an already published article, it is a *redundant publication*. The publication of a single result set as many articles is called *salami slicing.*

Such publications:

1. Waste the time of members of the scientific community
2. Unnecessarily expands the already extensive body of scientific publications
3. Infringes the copyright law
4. Confuses the scientific community by dispersing data taken from the same source
5. Overemphasises the significance of the research by repeated publications
6. Interferes with meta-analysis by boosting experimental or sample numbers

ii) Fabrication and falsification of data

Fabrication of data is the recording of fictitious data. Falsification is the manipulation of results or experimental details to make up a desired outcome.

According to a meta-analysis of surveys where scientists were asked if they had committed or know of a colleague who had committed academic scandal, an average of 1.97% confessed to falsifying or fabricating data and 33.7% admitted to having dubious research practices.[39]

Why do people falsify?[40]

i) Increased competition for highly sought after research budgets
ii) Tempting rewards for publishing in renowned journals
iii) Pressure of completing research to meet the expectations of peers or to meet a grant deadline

Implications of academic scandals[41]

i) Made up results can misled researchers, wasting time and money. It might even delay the development of crucial treatments or prolong the use of harmful drugs.
ii) Affects authors, reviewers and editors, but most importantly, it harms the patients
iii) Malpractice is easy to perpetuate but hard to reveal
iv) Erodes the trust of the public

Notable cases

In 2004, Professor Hwang Woo-Suk, a well-known South Korean doctor at the Seoul National University achieved international fame for his research on cloned human embryos. He amazed the scientific community by creating 11 patient specific stem cell lines.

However, in 2005, it was revealed that Hwang had falsified the data and that none of the DNA in the cell lines matched the original donor DNA. He lost his position and his papers on embryonic stem cell research were retracted from the journal *Science*.[42]

A lifetime of hard work and research can be brought down by a moment of folly. In order to perpetuate the trust of the public in the scientific community, scientists should occupy a moral high ground in the search for knowledge about nature. While ethics remain a highly debatable and controversial issue, the fundamental principle is to stand by what is morally sound and to do what will best benefit humanity.

References

1. Potter, S. (n.d.). BrainyQuote.com. Retrieved from http://www.brainyquote.com/quotes/quotes/p/potterstew390058.html
2. Smith, D. (2003). Five principles for research ethics. American Psychological Association. Retrieved from http://www.apa.org/monitor/jan03/principles.aspx
3. Resnik, D. B. (2011). What is ethics in research & why is it important? National Institute of Environmental Health Sciences. Retrieved from http://www.niehs.nih.gov/research/resources/bioethics/whatis/
4. Drew, C. J., Hardman, M. L. & Hosp, J. L. (2008). *Designing and Conducting Research in Education*. Europe: SAGE Publications, Inc., pp. 55–80.
5. U.S. National Library of Medicine. (2012). Greek medicine. Retrieved from http://www.nlm.nih.gov/hmd/greek/greek_oath.html
6. Wright, M. (2011). Ideals and the Hippocratic Oath. Retrieved from http://www.patient.co.uk/doctor/Ideals-and-the-Hippocratic-Oath.htm
7. Clados, M. S. (2012). Bioethics in international law: an analysis of the intertwining of bioethical and legal discourses. (Master's thesis). Retrieved from http://edoc.ub.uni-muenchen.de/15247/1/Clados_Mirjam_Sophia.pdf
8. World Medical Association. (2013). WMA Declaration of Geneva. Retrieved from http://www.wma.net/en/30publications/10policies/g1/
9. Rosenberg, J. (2014). Mengele's children — the twins of Auschwitz. Retrieved from http://history1900s.about.com/od/auschwitz/a/mengeletwins_2.htm
10. S. Government Printing Office. (1949). *Trials of War Criminals before the Nuremberg Military Tribunals under Control Council Law* No. 10, Vol. 2, Washington, D.C., pp. 181–182.
11. World Medical Association. (2014). WMA Declaration of Helsinki — ethical principles for medical research involving human subjects. Retrieved from http://www.wma.net/en/30publications/10policies/b3/

12. The National Commission for the Protection of Human Subjects of Biomedical and Behavioral Research, Department of Health, Education, and Welfare. (1979). The Belmont Report. Retrieved from http://www.nus.edu.sg/irb/Articles/Belmont Report.pdf
13. Medical Protection Society. (2013). Confidentiality — general principles. Retrieved from http://www.medicalprotection.org/singapore/factsheets/confidentiality-general-principles
14. Plaza, J. & Fischbach, R. (n.d.). Current issues in research ethics. Retrieved from http://ccnmtl.columbia.edu/projects/cire/pac/foundation/
15. Milligan, C. (n.d.). Anonymity and confidentiality. Retrieved from http://www.lancaster.ac.uk/researchethics/1-2-anonconf.html
16. Health Research Authority. (2013). What is section 251? Retrieved from http://www.hra.nhs.uk/about-the-hra/our-committees/section-251/what-is-section-251/
17. Thirumoorthy, T. & Peter, L. (2013). Consent in medical practice 3. Retrieved from http://sma.org.sg/UploadedImg/files/Publications — SMA News/4508/CMEP.pdf
18. Unite For Sight. (2013). Consent, privacy, and confidentiality. Retrieved from http://www.uniteforsight.org/research-course/module4
19. Cook, K. (2006). Facts about animal research. Retrieved from http://www.pro-test.org.uk/2006/03/facts-about-animal-research.html
20. Musculus, M. (2012). On animal consciousness. Retrieved from http://dogbehaviorscience.wordpress.com/2012/08/11/on-animal-consciousness/
21. ProCon. (2014). Should animals be used for scientific or commercial testing? Retrieved from http://animal-testing.procon.org/
22. Americans for Medical Progress. Animal research means medical progress. Retrieved from http://www.amprogress.org/animal-research-benefits
23. Hajar, R. (2011). Animal testing and medicine. *Heart Views* 12(1): 42.
24. Murnaghan, I. (2013). Using animals for testing: pros versus cons. Retrieved from http://www.aboutanimaltesting.co.uk/using-animals-testing-pros-versus-cons.html
25. British Broadcasting Corporation. (2014). The ethics of speciesism. Retrieved from http://www.bbc.co.uk/ethics/animals/rights/speciesism.shtml

26. Hepworth, A. (2010). Animal research: the ethics of animal experimentation. Retrieved from http://www.stanford.edu/group/hopes/cgi-bin/wordpress/2010/07/animal-research/
27. National Advisory Committee for Laboratory Animals Research. (2004). Guidelines on the care and use of animals for scientific purposes. Retrieved from http://www3.ntu.edu.sg/Research2/Grants Handbook/NACLAR-guide Lines.pdf
28. SingHealth Group. (2014). The Institutional Animal Care and Use Committee (IACUC). Retrieved from http://research.singhealth.com.sg/Pages/InstitutionalAnimalCareandUseCommittee.aspx
29. Christie-Pope, B. *Use of the cadaver lab at Cornell College.* Retrieved from http://people.cornellcollege.edu/bchristie-pope/CadaverLab/CadaverLab_Rules.html
30. Chin, D. (2009). Genetic engineering: Why so controversial? Retrieved from http://serendip.brynmawr.edu/exchange/node/4935
31. Lee, S. (2007). Scientists push the boundaries of human life. Retrieved from http://www.riorevuelto.org/site/ip/ventana.php?id_articulo=3126
32. Advani, P. (2005). Agriculture sector in India. Retrieved from http://ncw.nic.in/pdfreports/impact%20of%20wto%20women%20in%20agriculture.pdf
33. Om Organics. (2014). Genetic engineering/GMOs — controversy. Retrieved from http://www.omorganics.org/page.php?pageid=95
34. Church of Scotland. (2010). Moral and ethical issues in gene therapy. Retrieved from http://www.srtp.org.uk/srtp/view_article/moral_and_ethical_issues_gene_therapy
35. ThinkQuest. (2000). Genetic engineering debates. Retrieved from http://library.thinkquest.org/C004367/be10.shtml
36. EuroStemCell. (2011). Human embryonic stem cell research and ethics. Retrieved from http://www.eurostemcell.org/files/Human_ES_ethics_1.pdf
37. Pitak-Arnnop, P., Schouman, T., Bertrand, J. C. & Hervé, C. (2008). How to avoid research misconduct — recommendations for surgeons. *J. Chir.* **145**(6): 534–541.
38. University of Delaware. (2014). Code of conduct. Retrieved from http://www.udel.edu/stuguide/14-15/code.html

39. Jain, A. K. (2010). Ethical issues in scientific publication. *Indian J. Orthop.* **443**(3): 235–237.
40. Pritchard, M. S. (2006). Overly ambitious researchers — fabricating data. Retrieved from www.onlineethics.org/Education/precollege/scienceclass/sectone/chapt4/cs1.aspx
41. Jha, A. (2012). False positives: fraud and misconduct are threatening scientific research. Retrieved from http://www.theguardian.com/science/2012/sep/13/scientific-research-fraud-bad-practice
42. Emanuel, E. J., Grady, C., Crouch, R. A., Lie, R., Franklin, M. & Wendler, D. (2011). *The Oxford Textbook of Clinical Research Ethics.* Oxford: Oxford University Press, p. 792.

Chapter 8

Statistics for Clinical Research

Yiong Huak Chan

The Research Process[1]

Apart from getting the Ethics Committee's approval (before conduct of the study) and patients' consent (during conduct of study), Table 8.1 shows the stages of a research study that need to be addressed in detail before a credible and clinically relevant result could be obtained.

Table 8.1: Stages of a research process.

	Percentage of contribution to validity of clinical results obtained/%
Stage 1	
Proper study design (Epidemiological/ Randomised Controlled Trial)	30–40
Sample size calculations[2] (Precision/Power Calculations)	
Stage 2	
Conduct of study/data integrity (Garbage in Garbage out)	50–60
Stage 3	
Proper database setup/statistics	10–20

It is essential that Stages 1 and 2 be properly set up; otherwise, even with the help of a statistician to perform the analyses, the results obtained will not be valid!

Statistical Analysis

It is of utmost importance that every researcher understand that *statistical significance* is governed by sample size. The bigger the sample size, the more likely one would obtain a significant *p*-value. In any study, the *clinical significance* (which could not be manipulated, unless through bias) is the relevant outcome that one is interested in before looking at the *p*-value. It is through this clinical relevance that sample size is calculated!

We shall use a hypothetical study to illustrate the various statistical analyses that are typically performed. Detailed discussions of these techniques (with templates for SPSS analyses) are found in the statistical series published in the *Singapore Medical Journal*.[3-8]

In this chapter, we shall emphasise the presentation of univariate (unadjusted) and multivariate (adjusted) results using linear regression (quantitative outcome, e.g., QOL), logistic regression (qualitative outcome, e.g., limb loss) and Cox regression (time-to-event outcome, e.g., time to limb loss). A word of caution — the results are fabricated for discussion purposes.

The hypothetical study was designed to determine the predictive factors for limb loss of subjects with diabetic foot problems (DFPs). One hundred and fifty diabetic subjects were recruited. Table 8.2 shows the variables collected.

Quantitative Outcome: QOL (Linear Regression)

Table 8.3. lists the unadjusted and adjusted predictors for the continuous QOL outcome. There are two ways to interpret the findings. Firstly, if we are only interested in determining whether there were any differences in QOL between those with limb loss compared to those without, the only *p*-value we are interested in is for the 'Limb Loss' variable. From the unadjusted (univariate) results, those with limb loss had a significantly lower QOL score by 6.5 units (95% CI

Table 8.2: Study variables.

	Data type	Codings
Dependent variables		
Limb Loss	Qualitative	1 = Yes, 0 = No
QOL	Quantitative	
Time to Limb Loss	Quantitative	
Independent variables		
Age (years)	Quantitative	
Gender	Qualitative	1 = Male, 2 = Female
Ethnic	Qualitative	1 = Chinese, 2 = Indian, 3 = Malay, 4 = Others
Hypertension	Qualitative	1 = Yes, 0 = No
Stroke	Qualitative	1 = Yes, 0 = No
Smoking	Qualitative	1 = Yes, 0 = No
Obesity	Qualitative	1 = Yes, 0 = No
Ischaemia (ABI < 0.8)	Qualitative	1 = Yes, 0 = No
Gangrene	Qualitative	1 = Yes, 0 = No
Infection	Qualitative	1 = Yes, 0 = No
Ulcer	Qualitative	1 = Yes, 0 = No

3.1–9.9), $p < 0.001$, compared to those without limb loss. This difference between the two groups remained significant after adjusting for the other covariates (the adjusted or multivariate results) though now the difference is 3.3 (95% CI 1.0–5.6), with $p = 0.005$.

For linear regression, it is important to check that the tolerance of the variables is not small (near zero), a sign of multi-colinearity problem which distorts the estimates and p-values.[7]

The second way to interpret the results from Table 8.3 is that we are interested to find which variables affect the QOL score. In this situation, the p-values of all the variables will be of interest. Univariate results showed that subjects with limb loss, ischaemia and gangrene, and also older people were significant predictors for lower QOL. Since age is continuous, the interpretation will be for an increase of one year, QOL score will reduce by 0.16 (95% CI 0.02–0.29) units,

Table 8.3: Unadjusted and adjusted predictors for QOL.

	Unadjusted		Adjusted		
	B (95% CI)	p-value	B (95% CI)	p-value	Tolerance
Limb loss	−6.5 (−9.9 to −3.1)	<0.001	−3.3 (−5.6 to −1.0)	0.005	0.851
Age	−0.16 (−0.29 to −0.02)	0.021	0.02 (−0.07 to 0.11)	0.598	0.792
Females	−0.17 (−3.4 to 3.1)	0.919	0.76 (−1.5 to 3.0)	0.497	0.769
Indian	1.6 (−2.9 to 6.2)	0.486	1.4 (−1.5 to 4.3)	0.335	0.797
Malay	−1.8 (−5.5 to 1.8)	0.324	−0.7 (−3.0 to 1.7)	0.567	0.753
Other race	7.9 (−2.2 to 18.1)	0.126	7.6 (1.3 to 13.9)	0.018	0.927
Hypertension	−1.7 (−5.4 to 1.9)	0.349	−1.5 (−3.8 to 0.9)	0.227	0.865
Stroke	−5.3 (−10.8 to 0.3)	0.061	0.7 (−2.9 to 4.3)	0.707	0.861
Smoking	−1.3 (−5.5 to 3.0)	0.550	0.9 (−1.8 to 3.7)	0.494	0.854
Obesity	1.2 (−4.6 to 7.0)	0.692	1.5 (−2.2 to 5.2)	0.414	0.884
Ischaemia	−16.0 (−18.0 to −14.0)	<0.001	−15.5 (−17.6 to −13.4)	<0.001	0.840
Gangrene	−4.4 (−7.8 to −0.9)	0.013	0.7 (−2.9 to 4.2)	0.705	0.349
Infection	3.6 (−0.5 to 7.1)	0.060	1.9 (−1.5 to 5.5)	0.270	0.370
Ulcer	−1.5 (−5.2 to 2.1)	0.407	1.1 (−2.4 to 4.7)	0.530	0.383

Note: Race (ref) = Chinese.

with $p = 0.012$. Upon multivariate analysis, age and gangrene "lost" their significance, only limb loss and ischaemia remained significant. Other races compared to Chinese subjects had a significant higher QOL, 7.6 (95% CI 1.3–13.9) units, with $p = 0.018$.

From this example, we can see that the results from univariates, though giving us a good perspective of the findings, is not the final conclusion. The old school of thought of only including significant univariates into multivariates should be given careful consideration as we could miss out some significant findings (for example, other races in this case). The recommendation is to include all relevant variables collected into a multivariate analysis and remove multi-colinear variables if necessary, to get the final results.

Qualitative Outcome: Limb Loss

When the outcome is binary, logistic regression would be performed. In this case, the estimates obtained will be odds ratios (OR). An OR >1 means a likelihood of getting the outcome of interest. Table 8.4 shows the template of presenting the unadjusted and adjusted results. From the univariate results, significant risk predictors for limb loss were ischaemia (OR = 2.3; 95% CI 1.1–4.8; $p = 0.027$) and gangrene (OR = 3.7; 95% CI 1.8–7.9; $p = 0.001$). Interestingly, infection was a protective predictive for limb loss (OR = 0.22; 95% CI 0.1–0.6; $p = 0.003$). For the checking of multi-colinearity for logistic regression,[8] we do not have the luxury of using the tolerance (though some recommend using linear regression with limb loss as an outcome to make use of the tolerance facility to check for multi-colinearity before performing logistic regression). Upon multivariate analysis, none of the predictors were significant. One possible reason may be that the sample size of 150, in this case, was not sufficient for a multivariate analysis. Another possible explanation is that there were actually no predictors for limb loss of 48.9%, but this could be "rejected" from the clinical significance of 48.9% gangrene subjects had limb loss compared to 20.4% without gangrene. Since this study is exploratory, a step-wise logistic regression is relevant to be performed and both 'Gangrene' and 'Infection' were "picked up".

Table 8.4: Unadjusted and adjusted predictors for 'Limb Loss' (logistic regression).

	Limb Loss (n = 44)	No Limb Loss (n = 106)	Unadjusted OR (95% CI)	p-value	Adjusted OR (95% CI)	p-value
Age						
Mean (s.d.)	62.6 (12.1)	58.6 (12.1)	1.03 (0.99–1.1)	0.067	1.01 (0.97–1.04)	0.565
Gender						
Male	21 (27.6%)	55 (72.4%)	1.0		1.0	
Female	23 (31.1%)	51 (68.9%)	1.2 (0.6–2.4)	0.643	1.6 (0.7–3.7)	0.312
Race						
Chinese	19 (27.9%)	49 (72.1%)	1.0		1.0	
Indian	10 (38.5%)	16 (61.5%)	1.6 (0.6–4.2)	0.325	2.2 (0.8–6.7)	0.147
Malay	14 (26.9%)	38 (73.1%)	0.9 (0.4–2.1)	0.902	1.03 (0.4–2.6)	0.954
Others	1 (25.0%)	3 (75.0%)	0.9 (0.1–8.8)	0.899	0.9 (0.1–10.0)	0.920
Hypertension						
No	10 (25.0%)	30 (75.0%)	1.0		1.0	
Yes	34 (30.9%)	76 (69.1%)	1.3 (0.6–3.0)	0.483	1.1 (0.4–2.9)	0.796
Stroke						
No	37 (27.2%)	99 (72.8%)	1.0		1.0	
Yes	7 (50.0%)	7 (50.0%)	2.7 (0.9–8.1)	0.083	1.6 (0.4–5.7)	0.484
Smoking						
No	24 (27.6%)	89 (72.4%)	1.0		1.0	
Yes	10 (37.0%)	17 (63.0%)	1.5 (0.6–3.7)	0.334	1.4 (0.5–3.9)	0.553

			OR (95% CI)	p	OR (95% CI)	p
Obesity						
No	41 (29.9%)	96 (70.1%)	1.0		1.0	
Yes	3 (23.1%)	10 (75.9%)	0.7 (0.2–2.7)	0.606	0.8 (0.2–3.6)	0.782
Ischaemia						
No	14 (20.3%)	55 (79.7%)	1.0		1.0	
Yes	30 (37.0%)	51 (63.0%)	2.3 (1.1–4.8)	0.027	1.7 (0.7–4.0)	0.208
Gangrene						
No	21 (20.4%)	82 (79.6%)	1.0		1.0	
Yes	23 (48.9%)	24 (51.1%)	3.7 (1.8–7.9)	0.001	2.9 (0.7–11.5)	0.124
Infection						
No	39 (36.8%)	67 (63.2%)	1.0		1.0	
Yes	5 (11.4%)	39 (88.6%)	0.22 (0.1–0.6)	0.003	0.5 (0.1–2.3)	0.365
Ulcer						
No	32 (29.4%)	77 (70.6%)	1.0		1.0	
Yes	12 (29.3%)	29 (70.7%)	0.99 (0.5–2.2)	0.991	1.2 (0.3–5.0)	0.772

Note: OR = 1.0 is the reference category.

Time to Event Outcome: Time to Limb Loss (Cox Regression)

For survival analyses to be performed, the outcome is time to event with censored observations — "information" on the subject concerned. In this example, the time to event is 'Time to Limb Loss'; events will be subjects with limb loss due to DFP and the censored observations would be:

- Subject still has limb intact
- Subject loss to follow up or died
- Subject loses limb but not due to DFP

In survival analysis, we want to determine which predictor has a "shorter time to event". For example, would subjects with gangrene be more likely to have a shorter time to limb loss compared to those without? Kaplan Meier is the univariate analysis for survival and we can compare two survival curves using the log-rank test. Figure 8.1

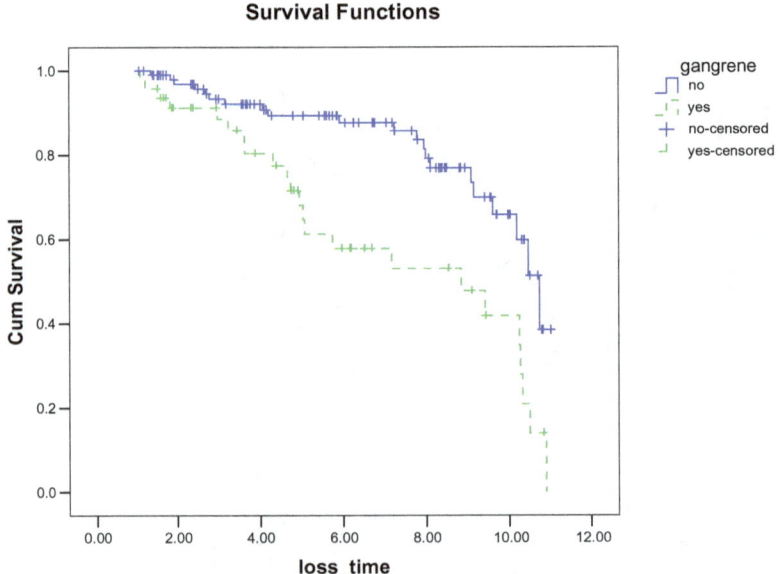

Fig. 8.1: Time to limb loss for gangrene.

shows the survival curves for those with gangrene compared to those without. The median time to limb loss for the gangrene group was 8.8 (95% CI 4.5–13.1) years compared to 10.7 (95% CI 10.0–11.4) years for those without gangrene, with $p = 0.001$ (log-rank test).

Table 8.5 shows the unadjusted and adjusted estimates for the outcome time to limb loss using Cox regression. The estimates

Table 8.5: Unadjusted and adjusted predictors for 'Time to Limb Loss'.

	Unadjusted		Adjusted	
	HR (95% CI)	p-value	HR (95% CI)	p-value
Age	1.03 (1.01–1.05)	0.038	1.01 (0.98–1.04)	0.583
Gender				
Male	1.0		1.0	
Female	1.2 (0.7–2.1)	0.571	1.8 (1.1–3.7)	0.042
Race				
Chinese	1.0		1.0	
Indian	1.3 (0.6–2.8)	0.500	1.8 (0.8–4.4)	0.173
Malay	0.9 (0.1–6.8)	0.712	1.1 (0.5–2.4)	0.765
Others	0.9 (0.4–1.8)	0.926	1.1 (0.1–8.5)	0.947
Hypertension				
No	1.0		1.0	
Yes	1.06 (0.5–2.1)	0.874	1.3 (0.6–2.8)	0.592
Stroke				
No	1.0		1.0	
Yes	1.5 (0.7–3.4)	0.335	1.03 (0.4–2.7)	0.953
Smoking				
No	1.0		1.0	
Yes	1.6 (0.9–3.3)	0.193	1.5 (0.6–3.3)	0.371
Obesity				
No	1.0		1.0	
Yes	0.5 (0.2–1.6)	0.243	0.5 (0.1–1.7)	0.250
Ischaemia				
No	1.0		1.0	
Yes	2.1 (1.1–3.9)	0.024	1.6 (0.7–3.3)	0.203

(Continued)

Table 8.5: *(Continued)*

	Unadjusted		Adjusted	
	HR (95% CI)	*p*-value	HR (95% CI)	*p*-value
Gangrene				
No	1.0		1.0	
Yes	2.7 (1.5–4.9)	0.001	2.8 (0.8–9.6)	0.093
Infection				
No	1.0		1.0	
Yes	0.3 (0.1–0.7)	0.009	0.6 (0.1–2.8)	0.561
Ulcer				
No	1.0		1.0	
Yes	0.99 (0.5–1.9)	0.981	1.3 (0.4–4.7)	0.681

Note: HR = 1 *is the reference category.*

obtained will be the hazard ratio (HR). An HR >1 means a likelihood to have a shorter time to event. From the multivariate results, the gangrene cohort, likely due to the small sample size, was not significant and females were at risk of having an earlier time to limb loss (HR = 1.8; 95% CI 1.1–3.7; $p = 0.042$). Figure 8.2a shows the limb loss survival times for both gender before adjustments; there is practically no difference. Upon taking the other variables into account, actually the females were "worse off" with more severe comorbidities (for example), thus being at risk (see Fig. 8.2b).

Conclusions

The above discussions on the various techniques have a coverage of at least 75–80% of analyses performed in published articles. For additional requirements, you may want to refer to the Refs. 10–18 or alternatively seek a consult from a statistician. It is also recommended to get a statistician involved in the planning stage of your study to assist in the Stages 1 and 2 of the research process before finally setting up the database and statistical analysis.

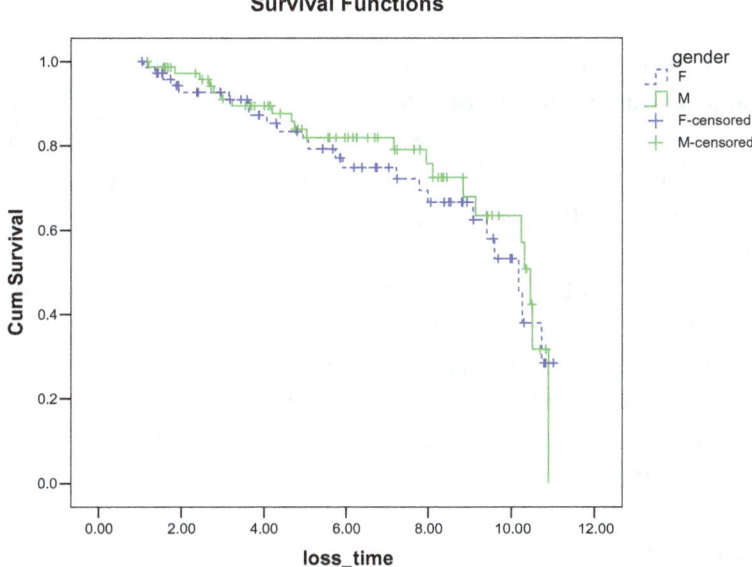

Fig. 8.2a: Time to limb loss by gender (unadjusted).

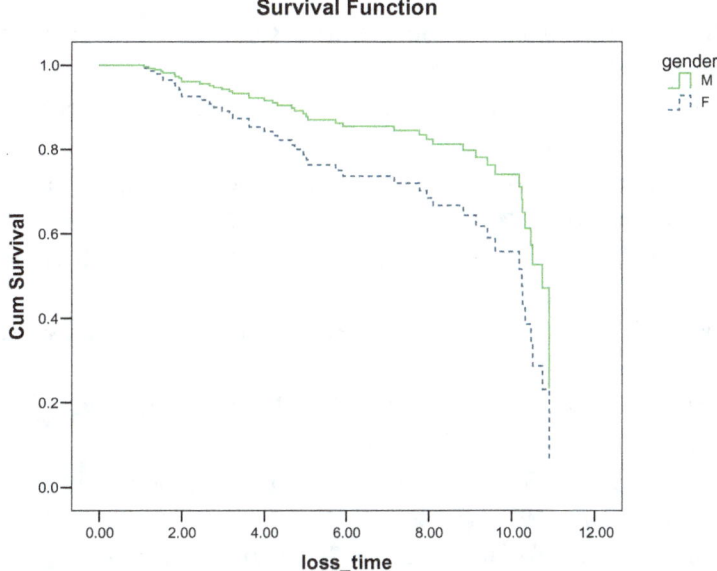

Fig. 8.2b: Time to limb loss by gender (adjusted).

References

1. Chan, Y. H. (2003). Randomised controlled trials (RCTs) — Essentials. *Singapore Med J.* **44**(2): 60–63.
2. Chan, Y. H. (2003). Randomised controlled trials (RCTs) — Sample size: The magic number? *Singapore Med J.* **44**(6): 172–174.
3. Chan, Y. H. (2003). Biostatistics 101: Data presentation. *Singapore Med J.* **44**(6): 280–285.
4. Chan, Y. H. (2003). Biostatistics 102: Quantitative data: parametric and non-parametric tests. *Singapore Med J.* **44**(8): 391–396.
5. Chan, Y. H. (2003). Biostatistics 103: Qualitative data: tests of independence. *Singapore Med J.* **44**(10): 498–503.
6. Chan, Y. H. (2003). Biostatistics 104: Correlational analysis. *Singapore Med J.* **44**(12): 614–619.
7. Chan, Y. H. (2004). Biostatistics 201: Linear regression analysis. *Singapore Med J.* **45**(2): 55–61.
8. Chan, Y. H. (2004). Biostatistics 202: Logistic regression analysis. *Singapore Med J.* **45**(4): 149–153.
9. Chan, Y. H. (2004). Biostatistics 203: Survival analysis. *Singapore Med J.* **45**(6): 249–256.
10. Chan, Y. H. (2004). Biostatistics 301: Repeated measurement analysis. *Singapore Med J.* **45**(8): 354–368.
11. Chan, Y. H. (2004). Biostatistics 301(a): Repeated measurement analysis (mixed models). *Singapore Med J.* **45**(10): 456–460.
12. Chan, Y. H. (2005). Biostatistics 302: Principal component and factor analysis. *Singapore Med J.* **45**(12): 558–566.
13. Chan, Y. H. (2005). Biostatistics 303: Discriminant analysis. *Singapore Med J.* **46**(2): 54–61.
14. Chan, Y. H. (2005). Biostatistics 304: Cluster analysis. *Singapore Med J.* **46**(4): 153–159.
15. Chan, Y. H. (2005). Biostatistics 305: Multinomial logistic regression. *Singapore Med J.* **46**(6): 259–268.
16. Chan, Y. H. (2005). Biostatistics 306: Log-linear models: Poisson regression. *Singapore Med J.* **46**(8): 377–385.
17. Chan, Y. H. (2005). Biostatistics 307: Conjoint analysis and canonical correlation. *Singapore Med J.* **46**(10): 514–517. Erratum in: **46**(12): 740 (2005).
18. Chan, Y. H. (2005). Biostatistics 308. Structural equation modeling. *Singapore Med J.* **46**(12): 675–679.

Section IV
Writing Your Research

Chapter 9

Tips for Scientific Writing

Claire Shu-Yi Chan, Wee Lin & Aziz Nather

Introduction to Scientific Writing

Writing is an art that needs to be mastered. Like any other art, the key to good writing lies in plenty of practice. Effective writing is key to writing a good research article. Good writing reflects good thinking on the part of the researcher. Moreover, research papers are more likely to be accepted if the English used is good.

The writing style of a scientific paper should be clear, concise and direct.[1,2] The aim of a scientific paper is to communicate information to the reader quickly and effectively. Several papers rely on complicated writing to appear academic. However, this style is often difficult for readers to understand. Writing in a straightforward and concise way will increase the readability and accessibility of your article. The following guidelines on scientific writing have been developed with this in mind. You might find them useful when developing your own scientific writing style. They have been broken up into eight simple rules for your reading pleasure.

Rule 1: Write an Outline First

Before embarking on your scientific paper, it is important to write an outline. The information to be presented should be organised in an appropriate order. There is often a considerable amount of data to be presented. A well-developed outline is useful in keeping ideas and information organised while writing. Outlines also help in the process of writing by acting as a conceptual skeleton for the paper to be built upon. A well-structured outline will also result in a paper with a good flow in your writing. This will be easy for the reader to follow and to understand.

Before Starting

The way to organise your paper depends on the type of paper you are writing. If you are writing an original article for a journal, it would be good to structure your information according to the standard IMRAD (Introduction, Methods, Results and Discussion) format. Other types of writing include meta-analyses, review articles, case series and case reports. These have their own writing formats as well.

The audience of your paper is also an important factor to consider before starting. How familiar is your audience with the research topic? If you are writing for a journal targeted for a more general audience (e.g., general orthopaedic surgeons for Clinical Orthopaedics and related research), the paper will require more explanation of technical terms and a general overview of the topic. You would write a different article when writing for a more specialised audience (e.g., to spine surgeons for the *Journal of Spine*).

Developing the Outline

Once you have decided on a suitable format for your paper, you can decide what information will go into each section of the paper.

List the major points to be included in each section, and arrange them in an order that makes sense. Also identify details to support each main point. To better illustrate some points, figures, diagrams, graphs, tables and charts can also be used. Remember "a picture is worth a thousand words". Some data is better and clearer when presented in a schematic diagram or graph. Certain facts and figures also need to be substantiated by supplying the correct references, which should be compiled beforehand.

Rule 2: Constructing Paragraphs

A research paper is written in continuous prose, which should be expressed in paragraphs. Knowing how much to include in each paragraph is essential, as appropriate paragraphing makes an article more readable and easy to follow.

Short and proper paragraphing is essential for a number of reasons:

- Makes one's paper easier to understand
- Ensures one's message is more clearly conveyed
- Avoids unnecessary detail and repetition
- Conveys clear description of one's argument

Rules for writing paragraphs:

1. One point per paragraph
   a. Each paragraph should be constructed to "tell a story". Keep to one point per paragraph. Do not cram too much information or too many details into a single paragraph — that will make it a long and tedious read. Breaking up your content into short paragraphs makes it easier for the reader to understand.
   b. Most readers are accustomed to the structure of one main idea in each paragraph. If you restrict yourself to one point per paragraph, the start of a new paragraph will easily signal to your reader the start of a new topic in your paper.
   c. Should you find your paragraph getting too long, split the content into two or more sub-points. Devote one paragraph to explaining each sub-point. This way, paragraphs can be kept short and to the point.

2. Topic sentences
   a. The topic sentence is a sentence introducing the main point. One should use a topic sentence at the beginning of each paragraph.

The topic sentence gives an overview of what is to be dealt with in the coming paragraph. Supporting sentences should also be well-organised in order for one to construct a logical argument.
   b. Placed at the start of the paragraph, topic sentences serve as a signpost for the reader. It informs the reader of the topic at hand, while also allowing the reader to anticipate the discussion that follows.
   c. Having a topic sentence at the start of each paragraph will also help to direct your writing. The rest of the paragraph should be organised to support the topic sentence.

3. Logical flow of information
   a. The topic sentence should be at the start of the paragraph to tell the reader what the paragraph is about.
   b. The paragraph body should support and flesh out the main point stated in the topic sentence.
   c. After finishing a paragraph, read through it to check that all sentences and information in it help to develop the main point. If there are any irrelevant sentences, delete them and place them in paragraphs where they are more relevant (or create another paragraph to elaborate on the new point).
   d. At the end of the paragraph, you can also reiterate the main point by restating the topic sentence. In doing so, the main point will be emphasised again before the reader moves on to the next paragraph.

Rule 3: Building Sentences

Sentences are the building blocks of paragraphs. They convey the ideas within a paragraph to the reader. How can you improve your sentences such that they communicate ideas with greater clarity?

Guidelines for sentence-writing:

1. Write short sentences
   a. One should strive to write in short sentences, as they are easier to understand than longer ones. In fact, contrary to popular

belief, the more difficult the science, the more important it is for one to keep the writing short and simple.

2. Subject–verb proximity
   a. The verb should follow immediately after the subject of the sentence. A common mistake is to interrupt the subject–verb complex with many distracting details. As a result, by the time the reader finds the verb at the end of the sentence, he or she may already have lost track of what the subject was.

> **Wrong way to write**
>
> **Investigation** of the differences, if any, between the socio-economic profiles of one cohort of patients presenting with diabetes mellitus and diabetic foot problems to the orthopaedic surgeon, and another cohort presenting with diabetes mellitus and no diabetic foot problems to the hospital endocrinologist, **was the objective** of this study.

   — The main subject of the sentence is *investigation* and the verb linked to it is *was the objective*. However, these two elements are separated by a large chunk of distracting detail.

   — This sentence is made clearer by rephrasing to maintain subject–verb proximity.

> **Better way to write**
>
> **The objective** of this study is to **investigate the differences** between the socio-economic profiles of one cohort of patients presenting with diabetes mellitus and diabetic foot problems, and another cohort presenting with diabetes mellitus without diabetic foot problems.

   b. Separation of the subject and verb can easily occur when too much information is crammed into a single sentence. Split long sentences into two or more parts using full stops. This will help to make sentences concise. Each sentence or clause delivers one piece of information, with the verb following directly behind the subject.

3. Placement of information
   a. Put known information first
      i. Known information is information that has previously been referred to, earlier on in the paper. Referring to known information at the start of a sentence is useful in two ways:
         - Firstly, it provides context for new information that will arrive at the end of the sentence.
         - Secondly, reiterating information helps a paper to flow more logically. You can make your sentences coherent by using known information to link new information to the rest of the paper.
   b. Put new information last
      i. New information should be placed at the end of the sentence, once the context is successfully established at the beginning. This position of emphasis at the end of a sentence helps to underscore the new information in the mind of the reader.
      ii. As such, this end position should be reserved for the most important new information in the sentence.
      iii. Putting new information at the start of the sentence introduces a new idea without providing any context. This is confusing for readers, who do not know how this new idea relates to the rest of the paper.

Rule 4: Use Proper Choice of Words

The word choice in a scientific paper reveals the writing style of the author. Choosing simple, direct and specific words is key to writing a scientific essay well. Listed below are some guidelines for word choice, as well as pitfalls of scientific writing that are best avoided.

1. Place actions in verbs
   a. Express the action of a sentence in the main verb, rather than using nominalisations. Sentences become easier to understand

when the verb follows directly after the subject. Using nominalisations instead of verbs makes this impossible.
b. Nominalisations are the noun forms of verbs. For example:
   i. *to analyse* → *analysis*
   ii. *to regulate* → *regulation*
   iii. *to correlate* → *correlation*
   iv. *to investigate* → *investigation*

> **Wrong way to write**
> **Excision** of the soft tissues and periosteum, as well as **removal** of marrow from the medullary canal of the segment, **was carried out**.

— This sentence contains two nominalisations: *removal* and *excision*. Here, the action of the sentence lies not in the verbs *to remove* and *to excise*, but instead in the *carrying out* of *removal* and *excision*. Here, nominalisations make the sentence unnecessarily long. The relationship between the subjects (*soft tissue and periosteum*, and *marrow*) and the actions performed on them is also blurred.

— Placing the actions in the main verbs instead of using their nominalisations makes this sentence more concise and direct.

> **Better way to write**
> The soft tissues and periosteum were **excised** and the marrow **removed** from the medullary canal of the segment.

c. Readers expect the action of a sentence to be explained by the verb. When it is not, readers will know what the subject of the sentence is, but not what actions it is performing — which is the crux of the sentence! To avoid confusion, place actions in verbs.

2. Wordiness
   a. Omit unnecessary words
      i. Unnecessary phrases which add no value
         - These are phrases which "pad" your sentences and make them look longer, but do not contribute any new information.
         - E.g., *Respectively, it should be noted that, etc.*
      ii. Longer phrases when shorter ones will do
         - These are lengthy phrases which can be replaced by shorter ones or even by a simple word.
         - Examples:

Lengthy phrase	Equivalent word
The vast majority	Most
Are capable of	Can
Due to the fact that	Since
With precision	Precisely
In the event of	If

3. Vague words and qualifiers
   Avoid vague words that provide only qualitative information to the reader. When writing a scientific paper, your audience is most likely interested in details of your materials and methods, and results. Qualitative words will not provide them with the useful information that they are looking for.
      i. Common qualitative words and phrases to avoid: quite, rather, several, fairly, sufficiently, appropriate
   b. Beware also of qualitative comparisons which ultimately do not tell your reader anything.
      i. Less, more, larger, etc.

4. Colloquialisms and contractions
   a. Colloquialisms specific to your field may not be known to the reader. These should be avoided in favour of more precise terminology. This will make the article better understood by readers who may not be familiar with the field.

Contractions such as *it's, there's* have no place in a scientific paper. Short forms such as *lab* should also be avoided.

5. Limit the use of jargons, abbreviations and acronyms
   As Wright *et al.*[1] pointed out, one should "avoid using jargon, acronyms and abbreviations" when writing. If one has to use acronyms and abbreviations, one has to ensure that that they are clearly defined the first time they appear in the text.

   Example of acronyms: DNA for deoxyribonucleic acid
   Examples of abbreviations: Dr for doctor

6. American and British spelling
   Is your journal American or British? Ensure you conform to the correct style of English required by the journal you are submitting your article to. If the journal does not have a specific preference, you can use either, but be consistent. If one is unsure of the spelling of a particular word, one should check the spelling using a good dictionary.

7. Know the meaning of every word
   a. One must clearly understand the meaning of all words used. Avoid using "big words" or words you are unsure of. If used wrongly, such words can lead to misunderstandings and confusion for readers.
   b. Remember that "big words" will not necessarily impress readers. Instead, readers may be confused if they do not understand the meaning of such words used.
   c. When in doubt, one should use a medical dictionary to clarify the meaning of a word. A thesaurus would also be useful for one to capture a wider vocabulary of words.
   d. One should avoid using complex words — settle for simpler words that are equally descriptive instead.
   e. The application of this simple concept not only increases accuracy of thought, but also eliminates many common grammatical mistakes. It ensures that the text becomes simpler and clearer to understand by both readers and reviewers.

Rule 5: Grammar and Punctuation

Good grammar and punctuation holds a scientific paper together. You may not notice the importance of good grammar, but poor grammar is immediately obvious. Not only is it irksome to read, it may also cause confusion for the reader who has to decipher what a poorly crafted sentence really means. A detailed discussion of grammar problems is out of the scope of this book. However, if your command of English requires further work, it is a good idea to look up grammar books or websites.

Guidelines for grammatical usage in scientific writing:

1. Use of the past/present tense
   a. When describing your own experimental work or that of others, the past tense should be used since you are speaking of events that happened in the past.

> ### Examples
> — We **found** that...
> — Smith *et al.* **provided** individualised patient teaching...
> — A detailed medical history **was taken** upon admission...

   b. When discussing your results, however, it is more appropriate to use the present tense as you are discussing your current thoughts and ideas.

> ### Examples
> — This study **is** an evaluation of...
> — Our results **show** a clear trend...
> — We **conclude** that...

2. Use of the third person
   a. The third person construction is the most accepted for scientific writing. It does not directly involve the writer (first person "I") or the reader (second person "you").

> **Examples**
>
> First person
>
> **I/We** documented data using a carefully designed study protocol. Treatment administered was determined by **us**.
>
> Second person
>
> **You** should use a carefully designed study protocol to document data. Treatment administered may be determined by **you** or **your** medical team.
>
> Third person
>
> **One** can use a carefully designed study protocol to document data. Treatment administered was determined by **the medical team**.

   b. This results in a more objective and professional tone.

3. Use of the active voice and passive voice
   The active voice increases one's clarity and the effectiveness of writing. On the other hand, the passive voice results in more words than necessary to convey the same message.
   a. Brief overview
      i. Most sentences have "actors" and "receivers of action".
         - E.g., **Nurses** educated the **patients and caregivers** before discharge.
         - Here, "nurses" are the actors who carry out the education, while "patients and caregivers" are the receivers of this action.
      ii. Active voice: the actors "act upon" the receivers of action directly.
         - Structure: **Actor**-<u>verb</u>-**receiver**
         - E.g., as above: **Nurses** <u>educated</u> the **patients and caregivers** before discharge.
      iii. Passive voice: the subject "is acted upon" by the verb
         - Structure: **Receiver**-<u>verb</u>-**(actor)**
         - E.g.,
            — The **patients and caregivers** were <u>educated</u> by **nurses** before discharge

— The **patients and caregivers** were <u>educated</u> before discharge.
- Note that the actor is optional; it can easily be omitted from the sentence
  b. Use active voice most of the time
     i. Missing actors in the passive voice:
        - Since the passive voice allows the actors to be "hidden", use of the passive voice can leave out important information about the actors. Readers may know the action and the receiver, but not know who carried out the action. If the actor is self-explanatory or unimportant, this does not matter. However, when there are multiple possibilities for the actor, using the passive voice leaves it uncertain.
     ii. Less wordiness with the active voice:
        - Sentences written in the active voice tend to be shorter. This is good, as a more concise paper is less time-consuming to read.
     iii. Thus, use active voice where possible. It is more direct and concise than the passive voice, and is often clearer for the reader.
     iv. E.g.,
        - Passive: "Oxygen was consumed by the rabbit at a regular rate."
        - Active: "The rabbit consumed oxygen at a regular rate."
  c. Use the passive voice to emphasise the receiver of action
     i. When you want to draw attention to the receiver
        - **All patients diagnosed with diabetic foot problems** were placed on the clinical pathway and included in the study.
        - Since the passive voice switches the position of the actor and receiver, you can use it to place the receiver at the beginning of the sentence for emphasis.
     ii. When the actor is not important
        - The one-way ANOVA was used to compare the mean hospitalisation cost from 2005 to 2010.

- Here, the actor (the researcher carrying out statistical tests) is of little relevance and is better omitted.
  — This is often the case when describing materials and methods, where the focus is on the techniques used rather than the researcher performing them.

4. Using computerised checking tools
   a. Spelling and grammar checkers
      i. These can be useful for a quick check of your paper. Computerised spelling checkers are especially useful for standardising American or British English in your paper — simply select the appropriate option for your spellchecker.
      ii. Manual proofreading of your paper is still necessary though. Spellcheckers may not pick up typographical errors that result in different (but incorrectly used) words.
         - E.g., for → four; moral → morale; peak → peek, etc.
   b. "Find and replace" tools
      i. Use "find and replace" to check for unnecessarily long phrases and replace them with their shorter equivalents.
      ii. Examples:

Lengthy phrase	Equivalent word
The vast majority	Most
Are capable of	Can
Due to the fact that	Since
With precision	Precisely
In the event of	If

Rule 6: Be Simple and Concise

By using simple language, one enforces accurate thinking. Conversely, the consequence of writing verbosely is that one often becomes inaccurate.

One should consistently ask oneself whether a sentence can be shortened, simplified or deleted. Every word used should serve a specific purpose. If it does not, it should be removed. As Strunk and White[2] espoused, "Vigorous writing is concise. A sentence should contain no unnecessary sentences, for the same reason that a drawing should have no unnecessary lines and a machine no unnecessary parts. This requires not that the writer makes all sentences short, or avoid all detail and treat subjects only in outline, but that every word tell."

In any case, excess words often divert the reader's attention and hinder his learning process. Furthermore, page limits often call for one to make the best of any available space.

E.g.,

Wrong: "It is important that we must write concisely with precision."

Correct: "Write precisely" or "Be precise"

Rule 7: Use of Specific and Concrete Language

Strunck and White[2] posited that "the surest way to arouse and hold the reader's attention is to be specific, definite and concrete". Concrete language helps to actively engage the reader by raising specific questions in his mind. This benefits in the reader's learning process. He learns as he seeks answers to the questions raised.

Specific language clearly points out essential details to the reader. It makes it easier for the reader to comprehend the broader implications of the issue at hand. It ensures that the reader firmly grasps and understands the core ideas of the paper. One should avoid using vague qualifiers such as "quite", "rather" and "several" in one's research paper. Instead, such qualifiers should be replaced with specific quantitative information that strongly substantiates one's case.

Rule 8: Checking the Flow of Your Scientific Paper

After you have finished writing your paper, give it a read-through to check that it has a good flow in the writing. A paper with good flow should read smoothly from one sentence to the next, and from one paragraph to the next. If you have planned your outline well, there should be a good sequence of ideas. Should your sentences or paragraphs seem to jump jerkily from one idea to another, you can use transitional words to make the relationship between successive sentences clearer.

Tip

Use transitional words such as "therefore", "because" and "although" between paragraphs. These words help to:

- Serve as powerful indicators of logical relationships
- Provide strong links between paragraphs

Some common transitional words:

- However/Nevertheless/In contrast/Conversely/Notwithstanding
- Therefore/Thus/Hence/Consequently/As a result/Accordingly
- Additionally/In addition/Also/Furthermore/Moreover

References

1. Wright, T. M., Buckwalter, J. A. & Hayes, W. C. (1999). Writing for the journal of orthopaedic research. *J. Bone Joint Surg.* **17**: 459–466.
2. William, Jr., S. & White, E. B. (2009). *The Elements of Style* (5th ed.). Boston: Allyn and Bacon.

Chapter 10

Choosing the Right Journal

Wee Lin & Aziz Nather

What are the Different Kinds of Publications that are Available?

There are a wide range of publications that one can choose to publish in:

- Traditionally printed journals
  There are now more traditionally printed journals than ever before. As such, there are many more opportunities for one to publish one's article. There are mainly two different types of journals: general or specialised journals. Clearly, general journals deal with a wider breadth of topics, while specialised journals are more specific in their area of focus. Moreover, journals also differ in their focus and areas of knowledge (such as academic journals, trade journals, etc.)
- Internet journals
  Many of the new internet journals are open access. This means that in order for one to publish one's paper, one has to pay the internet journal. The advantage of publishing on an internet journal is that when the author has paid a fee, there are no access costs for readers. In theory, this means that one's papers should be able to reach a wider audience as opposed to traditionally printed journals. (For

examples of a range of open access journals, visit the Biomed central website: www.biomedcentral.com.)
- Conference proceedings and posters
- Free Access Internet news groups

Free access internet news groups allow for one's ideas to be published and disseminated rapidly. However, as such articles are not indexed in computer databases, they are likely to vanish rapidly.
- Seminar proceedings
- Books
- Reports
- Informal newsletters and journals

Sources: http://api.ning.com/files/hSOUK3FODoubcKRzbPH7Wv7MNwDF6ZrsTc6tzLl8h8gM0UkjQPwD-kEYCkV7CFs5VF1vo3VrmH5OJdwUe5KsrLhMx4imhXa2f/InternationalJournalofHealthResearchand Innovation.jpg?crop=1%3A1&width=171, https://images.springer.com/sgw/journals/medium/40596.jpg, http://jenmccleary.com/wp-content/uploads/2011/12/Journal-of-Research-in-Childhood-Education.jpg

Nevertheless, it is advisable for academics to stick to journals (both internet and printed) for publishing.

Why is it Important to Choose the Right Journal?

The objective assessment of the research output of an individual hinges on:[1]

- Publication in journals
- Book chapters/review articles

- Theses/monographs
- Research grants obtained
- National/international awards
- Membership on national research committees/editorial boards
- Peer review

> **Take note!**
> It would be good for one to begin deciding on a suitable journal as soon as one commences on one's project, as this helps to streamline the process.

Among these factors, publication output is the most discerning of an individual's research output and is critical to individual promotion. In fact, the success of one's research is best judged by the number of Tier 1 or Tier 2 international refereed publications produced.

As have been previously dealt with, there are a wide range of journals available. Each journal specialises in a highly specific area of research. Consequently, the readership varies considerably from one journal to the next. As such, ensuring that one chooses a suitable journal is paramount as it affects the impact and readership of one's research article.

In order for one's research article to make a lasting impact, one must be clear of one's intended audience, and choose a journal that would allow it to best reach out to its intended audience.

What are the Key Considerations to Take into Account when Deciding on a Journal?

There are a few considerations one should bear in mind before deciding on a journal.

> **Important to note**
> The higher the ranking of the journal, the wider its reach and the greater its impact. However, this also means that it is more difficult to get one's paper published in a journal of higher ranking. Nevertheless, one should **aim high**. If one aims for a journal with a high citation ranking, and does not meet its qualifications, one can still **revise one's target journal** to a lower ranked one later.

Current Updated Ranking of Journals

One should look at the current, updated rankings of journals (such as the NUS ranking of journals). In general, peer-reviewed journals tend to rank higher than journals that have not been peer-reviewed.[1]

There are three important criteria that determine a journal's ranking: absolute citation frequency, immediacy index and impact factor. The higher the ranking of the journal, the wider its reach (based on impact factors, absolute citation frequency, immediacy index[1]) and the more likely one's research would have a greater impact.

Absolute citation frequency[2]

The absolute citation frequency refers to the number of times a particular journal has been cited in other journals. The disadvantage that lies therein in this method is that a journal published more frequently theoretically has an advantage over one that is published less frequently. As such, this measure may not provide an accurate gauge of the ranking of the journal.

Immediacy index[2]

The immediacy index is a measure of how quickly an article in a particular journal is cited. It can be calculated by taking the number of articles from a particular journal that have been cited in a given year, divided by the number of all articles published by the same journal in the given year. Once again, this measure runs into problems as it favours journals which are published more frequently, hence giving rise to a biased ranking of journals. After all, a journal published more frequently would have a higher probability of being cited than one published less frequently.

Impact factor[2]

The impact factor can be taken to be a relative approximation of the importance of a journal within its field of study. The higher the impact factor of the journal, the more important it is deemed to be.

The impact factor is determined by measuring the average number of times articles from a particular journal are cited. It can be roughly calculated by taking the number of all the citations published in a given year by the journal in question, divided by the number of all articles published by said journal in the chosen years.

By far, the impact factor is the most reliable index when it comes to the ranking of journals. This is because the impact factor is not affected by the size of the journal, frequency of publication of the journal and age of the journal. The impact factor of journals is published in annual volumes of the *Journal Citation Reports*, and in various citation indices published by the Institute of Scientific Information.

The National University of Singapore (NUS) Tiering List 2012 Ranking Exercise serves as a useful guide for research workers. Using impact factor as its only consideration, journals are ranked according to their various specialties (such as orthopaedic surgery, trauma) and sub-specialties (such as spine, hip and knee, sports medicine, etc.). Journals are ranked into five different tiers. The higher up the tier a journal is ranked, the better regarded it is.

Tier I: Top 10% of all journals graded

Tier 2: Top 20% of all journals graded

Tier 4: Top 25% of all journals graded

Tier 4: Top 45% of all journals graded

Untiered: Remaining 55% of the journals graded

With all this factors taken into account, one should preferentially choose a journal that has been internationally refereed to be Tier 1 or Tier 2, with a high impact factor.

Standard of One's Research

The journal chosen should be tailored to the standard of one's research. There are mainly three different types of journals: international refereed, regional refereed and local refereed. Below is a brief list of journals belonging to each category.

International refereed: *Journal Bone and Joint Surgery* (American), *Journal Bone and Joint Surgery* (British), *Clinical Orthopaedics and Related Research*.

Regional refereed: *Journal of Orthopaedic Surgery, Journal of ASEAN Orthopaedic Association, Malaysian Orthopaedic Journal* (the *Malaysian Orthopaedic Journal* is a peer-reviewed journal accepted by West Pacific Region Index Medicus. However, it has yet to secure a PUBMED ranking.).

Local refereed: *Annals Academy of Medicine.*

Internationally peer-reviewed journals are regarded more highly than regional refereed journals. The latter is in turn better received than local peer-reviewed journals.

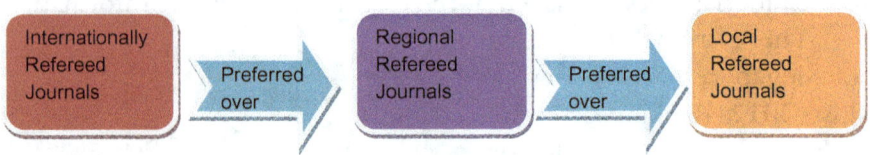

Suitability of Journal to One's Research

Last but not least, one should consider how suited the journal is for one's research article. The journal chosen should be relevant to one's area of research. With knowledge becoming increasingly specialised, it would be good for one to submit articles to journals that are geared towards one's relevant subspecialty. A journal which focuses more on one's chosen research area would inevitably be more suited for one's article than a journal whose focus is only loosely connected to one's field of research.

Journals suitable for clinical research are grouped under General Orthopaedics (Table 10.1), Spine Surgery (Table 10.2), Adult Reconstruction (Table 10.3), Trauma Surgery (Table 10.4), Ankle and Foot Surgery (Table 10.5), Hand and Reconstructive Surgery (Table 10.6), Shoulder and Elbow Surgery (Table 10.7) and Tissue Engineering (Table 10.8). Journals suitable for Basic Research are listed in Table 10.9. The impact factors for these journals are listed for the year 2012.

Table 10.1: General orthopaedics. (Based on 2012 impact factor)

Title of journals	Impact factor
Journal of Bone and Joint Surgery — Am Vol	3.23
Journal of the American Academy of Orthopaedic Surgeons	2.46
Journal of Bone and Joint Surgery — Br Vol	2.69
Clinical Orthopaedics and Related Research	2.79
Journal of Orthopaedics and Sports Physical Therapy	2.95
Acta Orthopaedica	2.74
International Orthopaedics	2.32
Orthopedics	1.05
Journal of Orthopaedic Surgery and Research	1.01
European Journal of Orthopaedic Surgery and Traumatology	0.181

Table 10.2: Spine surgery.

Title of journals	Impact factor
Spine	2.16
European Spine Journal	2.13
Journal of Spinal Disorders and Techniques	1.77
The Spine Journal	3.36

Table 10.3: Adult reconstruction.

Title of journals	Impact factor
American Journal of Sports Medicine	4.44
Gait and Posture	1.97
Journal of Arthroplasty	2.11
Operative Techniques in Sport Medicine	0.182

Table 10.4: Trauma surgery.

Title of journals	Impact factor
Journal of Orthopaedic Trauma	1.75

Table 10.5: Ankle and foot surgery.

Title of journals	Impact factor
Foot and Ankle International	1.47
Journal of the American Podiatric Medical Association	0.768
Journal of Foot and Ankle Surgery	0.860

Table 10.6: Hand and reconstructive surgery.

Title of journals	Impact factor
Journal of Hand Surgery — Am Vol	1.57
Hand Clinics	0.946

Table 10.7: Shoulder and elbow surgery.

Title of journals	Impact factor
Journal of Shoulder and Elbow Surgery	2.32

Table 10.8: Tissue engineering.

Title of journals	Impact factor
Tissue Engineering	4.07

Table 10.9: Basic research.

Rank	Title of journals	Impact factor
Tier 1	Bone	3.82
Tier 1	Journal of Orthopaedic Research	2.88
Tier 1	Calcified Tissue International	2.50
Tier 2	Clinical Biomechanics	1.87
Tier 3	Connective Tissue Research	1.79
Tier 4	Journal of Orthopaedic Science	0.96

How do you Know What is the Right Journal for you?

Since there are a huge range of journals one can potentially choose from, it is crucial that one first narrows down the possibilities by understanding the focus and readership of the different journals. One should compare the focus and readership of the journals chosen with one's intended target audience, and the content of one's article.

One should also look through past issues of the journals in question. In doing so, one should ascertain whether one's article is suitable for the readership of the journal. A question that one should constantly ask oneself is: does

> **Remember to**
>
> Reference articles from the journal chosen in one's research article. This demonstrates the relevance of one's article and its propensity for impact.

this journal allow people interested in similar topics to find out more about one's study? A good indication that a journal is suited for one's article would be if there were studies similar to one's article that were published in previous issues of the journal in question.

Reading through previous editions of the journals allows one to familiarise oneself with the format of the journal in question. One should also determine the type of articles published by the journal — whether the journal predominantly publishes case reports, review articles or original articles. Original articles should be awarded the most points, and case reports the least. Review articles should be awarded lower points compared to original articles. This reflects scholarship of integration, another component of intellectual activity of the individual (Holmes *et al.* 2000).[3]

Other journal publications include editorials, further opinions and letters to editors. Book chapters can also be counted as publications. More points should be awarded if an entire book is written by a sole author. Conversely, fewer points should be given for editing a book with only several chapters contributed by the editor. Even fewer points are to be allocated for editing a book without any chapter contributed by the editor. Additionally, lower recognition is given to a book chapter than to original articles in peer-reviewed journals. This is mainly because books are not peer-reviewed. However, if the book goes on to become a bestseller, the book becomes more important as a publication since the market reviews the book.

Having understood the focus and readership of each journal, and glimpsed through past issues of the journals in question, it would then be possible for one to narrow the scope. One should select two or three journals in the relevant area of research which have relatively high impact factors, and which are focused towards one's area of research.

Upon settling on two or three journals, one should then discuss with one's mentor on the possible journal options, and settle for the journal which can best capture the main essence and spirit of one's article. Furthermore, there are several online journal selectors such as Springer Journal Selecter (online app) that can offer an effective and fast search for journals best suited to one's research paper.

What are the Steps to Take Once you Have Decided on the Right Journal?

Once you have decided on the right journal for you, the next step that follows would be to check the journal's submission criteria and required format. Read the "Instructions to Authors" provided for by the chosen journal very carefully. Modify one's paper accordingly, ensuring that it is in compliance with the journal's requirements. Once this is done, one can then submit the paper to the journal!

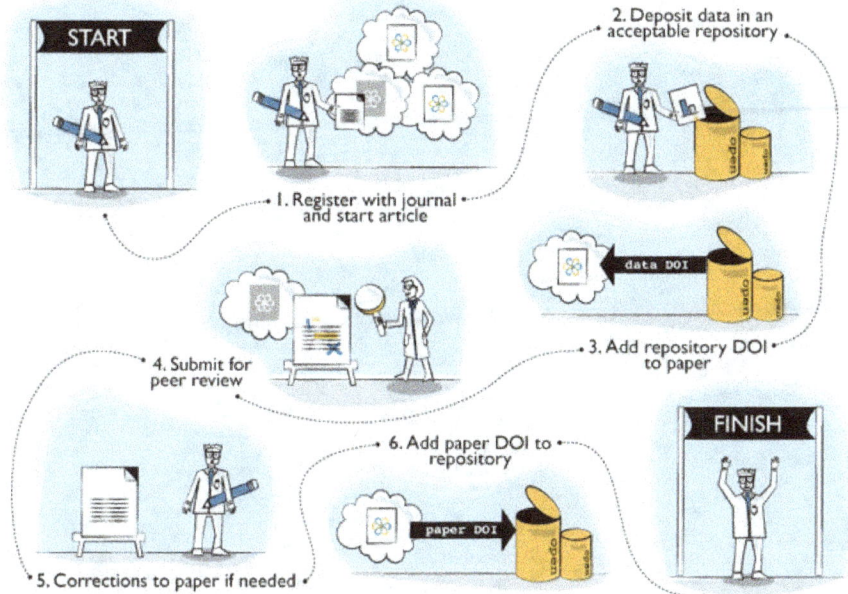

Source: http://openhealthdata.metajnl.com/about/submissions

References

1. Garfield, E. & Cawkell, A. E. (1975). Citation analysis studies. *Science* **189**(4200): 397.
2. Garfield, E., Malin, M. V. & Small, H. (1978). Citation data as science indicators. In: Elkana, Y., Lederberg, J., Merton, R. K., Thackray, A. & Zuckerman, H. (eds.). *Toward a metric of science: The Advent of Science Indicators*. New York: John Wiley & Sons, pp. 179–207.
3. Holmes, E. W., Burks, T. F., Dzau, V., Hindery, M. A., Jones, R. F., Kaye, C. I., Kom, D., Limbird, L. E., Marchase, R. B., Perlmutter, R., Sanfilippo, F. & Strom, B. I. (2000). Measuring contributions to the research mission of medical schools. Acad. med. 75: 304–313.

Chapter 11

How to Write an Original Research Article for a Journal

Wee Lin & Aziz Nather

"By the time I am nearing the end of a story, the first part will have been reread and altered and corrected at least one hundred and fifty times. I am suspicious of both facility and speed. Good writing is essentially rewriting. I am positive of this."

— Roald Dahl,[1]
British novelist and short story writer

Why Write a Research Article?

Why do Research?

Before one begins writing the research article proper, it is essential that one must first fully comprehend its purposes. Research is conducted for the betterment of society and for the advancement of Mankind. Jon Turner[2] rightly noted that the purpose of research is to "contribute to knowledge and society" and to "teach others what we have learned so they can use and build on our work".

Why is a Research Article so Important?

A research article is important because it conveys an idea and the results of one's research effectively and efficiently, from the researcher's mind to the reader's mind. This allows for the results of one's research to be made public, and for it to be used in beneficial ways.

Before You Start Writing ... Things You Need to Know

10 characteristics of an incredibly dull paper by Sand-Jensen[3]

1. Avoid focus
2. Avoid originality and personality
3. Make the article really, really long
4. Do not indicate any potential implications
5. Leave out illustrations (... too much effort to draw a sensible drawing)
6. Omit necessary steps of reasoning
7. Use abbreviations and technical terms that only specialists in the field can understand
8. Make it sound too serious with no significant discussion
9. Focus only on statistics
10. Support every statement with a reference

Knowing your Target Audience

It is essential that one understands the needs of the audience one is trying to reach out to, and keep them in mind while one is writing.

Writing for the General Practitioners/Primary Health Givers

If one intends to write a paper for general practitioners, one cannot assume that the reader already has thorough knowledge of the research area beforehand. Rather, one should presume that the reader does not know anything at all about one's topic. Bearing this point in mind, one should avoid any technicalities, and strive to explain things as simply to the layman as possible. All terms should be defined clearly, and relevant background information should be provided where necessary.

Writing for a Pool of Specialists

If one's target audience is a pool of specialists, one can then afford to use more technical terms. Even then, specialists still need to be informed of the specifics of what one has discovered in one's research. Still, the need to write simply and concisely remains.[4]

Handling the Competition

Competition for publishing one's research article in a journal is extremely stiff. While it is tempting for one to discredit and denounce the work carried out by one's rivals, this is neither smart nor necessary. Indeed, bad-mouthing the competition often proves to be injurious to one's paper:

- Being overly-critical may reinforce the idea that one is insecure and defensive.
- There are times where one's competition may happen to be the person reviewing one's paper.

Therefore, rather than making one's competition look bad, it would be advisable that one focuses on how one has built up on work that others have performed before.[2]

Choosing the Right Journal

The research paper should be targeted to be written for a highly rated (by the university involved) Tier 1 or Tier 2 internationally refereed journal of outstanding repute, with a high impact factor to gain maximal impact from the paper published. Each university has its own rankings of journals. It is wise to closely follow the university's order.

Consideration must also be made whether the work should be directed to a journal publishing General Orthopaedics (Table 11.1) and if the research embarked upon is more suited to a specialty, e.g., Spine Surgery (Table 11.2) or Tissue Engineering (Table 11.3).

On the other hand, if the work is dealing with an Orthopaedic Subspecialty, the top journal in that specialty should be chosen.

One must aim high and approach the highest tier journal that one estimates one's work could potentially be accepted. In the event the paper is rejected, the reviewer's comments can be studies and the writing re-directed to the next appropriate journal.

The aim is to get the paper published in an internationally refereed highly tiered journal. It is only when this fails that one could think of sending one's article to a regional or locally refereed journal.

Table 11.1: General orthopaedics.

Journal	Impact factor (Based on 2012)
Journal of Bone and Joint Surgery — Am Vol	3.23
Clinical Orthopaedics and Related Research	2.79
Journal of Bone and Joint Surgery — Br Vol	2.69

Table 11.2: Spine surgery.

Journal	Impact factor (Based on 2012)
The Spine Journal	3.36
European Spine Journal	2.133
Journal of Neurosurgery Spine	1.61

Table 11.3: Tissue engineering.

Journal	Impact factor (Based on 2012)
Tissue Engineering	4.07
Journal of Tissue Engineering and Regenerative Medicine	2.83
Journal of Tissue Science and Engineering	2.17

Authorship and Copyright Transfer

With a general guide on how to write an effective research paper in place, one can now begin the writing process. Prior to writing the research article however, there are several important issues one still has to address.

Firstly, it is vital that all investigators concerned meticulously read and comprehend the "Instructions to Authors" asked for by the journal chosen.

Secondly, it is essential for one to settle the issue of authorship. One must determine who should be the main author for the research article. The main author is not only the key investigator during the research process, but also shoulders the bulk of the responsibility for the written article. One also has to select co-authors who would assist the main author in both the research and writing process. It is also crucial that one discerns the order in which the co-authors are to be listed in the article. This is often dependent on the amount that each co-author has contributed to the research process. In doing so, one ensures that all who have contributed to the research and writing process are duly recognised. In most cases, supervisors or mentors of research projects are credited as co-authors. In cases where this is not so, supervisors should be recognised in the "Acknowledgements" section (elaborated below).

Finally, one can then complete the "Letter of Transmittal" which has to be signed by all authors. This letter is sent to the editor of the chosen journal.[5]

Format for Research Article

The format of a research article is:

- Abstract
- Introduction
- Materials and Methods
- Results
- Discussion
- Conclusion

- Acknowledgement
- References

Title

Before one reads the research paper proper, the first thing that captures one's attention is the title of the research article. Therefore, having a simple and concise title that effectively captures the main ideas of the article while catching the eye of the reader is vital. In other words, the title must be able to accurately reflect the significance of one's project while sustaining the reader's interest. For example, the title 'Research Methodology in Orthopaedics, Hand & Reconstructive Microsurgery' is not catchy enough a title, and can be replaced with a better and more concise title such as 'How to Do Research'.

There are a few points that must be captured within the title of the article:

- Subject matter
- Research results
- Authorship and the order of co-workers

In order to come up with an effective title, the title should be specific in describing the contents of the paper. It should not be too technical that only experts within the relevant field would be able to understand. This ensures that the paper reaches a wider audience, and does not put off people who may not understand the technical jargon.

Abstract

The abstract appears right at the beginning of the research paper, and provides a brief overview of one's study. It is typically about 200 to 350 words. The abstract can be "structured" or "non-structured". A structured abstract has sub-headings which summarises and accurately reflects the:

- Introduction
- Specific Objectives
- Clinical Significance

- Methodology
- Results
- Conclusion

Where the journal does not require a structured abstract, it is still wise to write the abstract along these lines, with one paragraph for each item.

Kent Beck[6] offered a general guide to writing an abstract. He enunciated that one should first state the problem, before explaining why it is clinically important. Following which, one should describe, in brevity, what one's solution achieves and what follows from it.

Like how a movie trailer serves to generate viewer interest in the movie, the abstract acts as a teaser as to what is to come in the later segments of the research paper. Therefore, the abstract is pivotal in determining reviewer and reader interest in subsequent parts of the article. Furthermore, the reviewer often uses the abstract as an indication as to whether a piece of work has potential or not. It is therefore of utmost importance that one writes a good and engaging abstract:

- Keep the abstract short and effective
- Exercise creativity in writing the abstract; this helps generate reader interest
- Do not include abbreviations, footnotes or citations in the abstract

Introduction

"Funnel Concept"

In many ways, the introduction is similar to that of the conclusion, except that it is written in reverse. The introduction typically consists of 2 to 4 paragraphs (about 250 words, usually comprising 20% of total length of study and not exceeding one A4 page) which:

- Provide a general background of the study
- Discuss previous work that has been done in the area of study
- Describe the problem one is addressing in the paper

- Elaborate on one's contributions to the area of research
- Raise new issues that are going to be addressed
- Justifies research relevance

The introduction starts first by giving a sketch of the general topic, before focusing on the specific area of research and the thesis statement. It is essential, however, that in the introduction, one does not repeat exact phrases used in the conclusion.

Hence, the introduction should provide a summary of what is already established and "known" in the area of study. It also identifies what is still "unknown" in the field of study. It then focuses on which part of the "unknown" the author intends to research on. The introduction ends with the specific objectives to be addressed by the project.

It is important to ensure that the introduction is well written, in order to motivate the reader to continue reading the rest of the research article. An example can be used to help introduce the problem. This not only illustrates one's point clearly, but also serves to make the introduction more interesting and engaging.

Materials and Methods

The experimental section should elaborate on the materials and methods one has used to arrive at the research results. This section should clearly and completely described, in chronological order, how the research was performed. This includes:

- Experimental design
- Statistical methods used to analyse the data
- Details of materials or instruments used

Source: http://www.toonpool.com/user/250/files/the_chemical_experiment_75405.jpg

- Names and locations of companies that have supplied various equipment and instruments for conducting one's research. This enables other investigators to reproduce the experiments.
- Relevant ethical considerations
- Preliminary results of experiments conducted prior to the main experiment, if it helps to explain the main procedure

> **Tip**
>
> If one's procedure is complicated, it may help to illustrate the research process with flowcharts, diagrams or tables.

However, it is important to note that the results of one's experiments should not be included at this point.

Results

In this segment, one should record in detail, the results attained through the course of one's experimental research. The order in which the results are presented should follow the discussion themes, rather than the order which they were attained through the experimentation process.[4]

Well-constructed tables and well-organised graphs often provide a useful pictorial representation of the results attained (for e.g., see Fig. 11.1). Clear pictures and illustrations are also useful to provide a

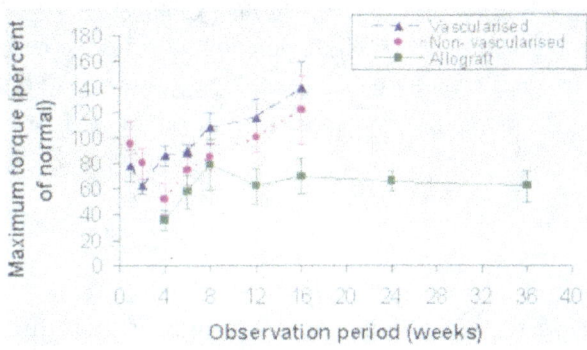

Fig. 11.1: Graph showing maximum torque of vascularised autograft, non-vascularised autograft and allograft in adult cats.

good and lasting impact. However, as the renowned Irish poet and writer Oscar Wilde remarked, "Everything in moderation, including moderation."[7] Hence, while pictures and graphs are useful as a visual portrayal of one's research results, they should not be used excessively, and should be limited to about three figures or tables. Main findings, nevertheless, should still be provided in the main text, in addition to the graphs and diagrams used.

It is also imperative to understand that charts and diagrams should only be used to illustrate points, rather than for aesthetic purposes. If a particular diagram does not play any role in helping to illustrate one's results, it should be deleted.

In this section, one can also consider comparing and contrasting one's results with the results of similar experiments conducted by others previously. In making such comparisons, it is crucial to bear in mind that one has to be objective. One should not exaggerate what one has done, but strive to make measured claims. It is essential that one acknowledges the weaknesses in one's approach, while giving credit to the competition when credit is due.

General tips on the use of diagrams

- Text used on charts should not be smaller than that used in the main body
- Do not use too many curves on a single chart as it tends to distract the reader's attention and obscures the message
- Use labels rather than legends, as the latter forces readers to refer back and forth and wastes the reader's time[2]

Discussion

The discussion section should aim to fulfil four main functions:

- **Answer the research questions and objectives previously put forth in the introduction section**

The discussion requires one to highlight the most significant results obtained during the research process. Notwithstanding, one should take measures to ensure that one does not simply regurgitate data which has already been presented in the "Results" section. Rather, one should attempt to explain how the results obtained answers the research question at the beginning of the paper. This should be done in reference to key data in figures and tables to remind the reader how answers were drawn from data.

A possible way to begin this section is to rephrase the research question and objectives, followed by answering the question using the results arrived at through the course of research. The answer should be stated explicitly and directly, to ensure that no ambiguity is implied.

- **Take into account the strengths and limitations of the present research study**
- **Consider whether the data obtained are consistent with one's hypothesis, and with what other investigators have reported in previous studies**

If one's results do not cohere with one's hypothesis, one can then attempt to explain why this may be so (for example, there may have been a different way of interpreting the results). One should also address the issues of how one's results fit in with findings already established, and whether any further research is necessary to address questions raised by the current results.

- **Conclude with a clear statement of the clinical implications and potential applications that may result from one's research**

At the same time, one should be careful not to speculate or extrapolate from the results obtained. Do not end the section with a statement of future studies that one may wish to pursue. Future plans are not relevant in the discussion of the article. Rather, one should end the discussion with a paragraph crystallising the answer the paper has provided, and its impact on clinical significance, if any.

At this point, it is vital that one asks oneself whether one's results has led to any new information, or merely serves to reinforce pre-existing knowledge on the topic.

If one's research has lead to new and promising results ...

The discussion should clearly and completely state how important this original work is in contributing to the pool of existing knowledge.

If one's research does not lead to any positive results ...

It is still useful to publish negative results so that others are informed of the study. To achieve (gain) publication, it is important that a good discussion must be written with a good review of literature.

In writing the discussion, one should focus on explaining how one's present work is different from that published by other research workers. Relevant points to be discussed include:

- A larger (bigger) cohort size
- A better methodology
- The use of animals higher up on the evolutionary scale
- More accurate experimentation equipment
- More data collected
- Stronger evidence in support of the conclusions reached

A good literature review must include articles from Tier 1 or Tier 2 refereed journals, and work from prominent institutions and well-known research workers. Randomised control studies and meta-analyses and should be preferred to case studies and case reports. References should also include the latest and most important contributions.

It is also important to include one or two references from the same journal the article is being sent to. The number of references to be appended depends on the type of journal it is being sent to. British journals usually prefer less than 10 good references, while a more extensive list of references is normally published in articles appearing in American journals.

The discussion is often the main factor that leads to the acceptance or rejection of a research article. While the methodology and results can be likened to the "body and limbs of the paper", the discussion can be said to be the "heart and soul of the paper". As such, the discussion written is extremely important in determining the success of a research paper:

- **The discussion should have good depth, to support the case in point.**
- **The paragraphs must be expressed in the correct sequence to ensure a good flow in the line of argument.**
- **It should include a good and extensive review of literature.**

Source: http://blog.journals.cambridge.org/wp-content/uploads/2012/06/shutterstock_9558

Conclusion

The conclusion should aim to answer the question that is being asked in the introduction. It should include:

- Significant results of the study
- Clinical significance of the study
- Possible future applications of this work

Acknowledgements

It is important to thank all who have contributed in one way or another throughout the process of one's research. One should thank the funding agency, and all colleagues, scientists and technicians who have rendered assistance.

"Cultivate the habit of being grateful for every good thing that comes to you, and to give thanks continuously. And because all things have contributed to your advancement, you should include all things in your gratitude."

— Ralph Waldo Emerson, American poet[8]

Giving credit to another does not diminish the credit one gets from the paper. One should warmly acknowledge all who have helped. In fact, failing to give credit to others can be detrimental to one's paper. It amounts to plagiarism.

References

In order to avoid plagiarising another's work, one should remember to cite all references that have been used over the course of one's research and writing process.

Moreover, referencing the work of others has the advantage of allowing other interested researchers to find information on resources referred to in the text. It also lends greater support and validity to the author's argument by showing that he has done the appropriate literature review in the field of work studied.

There are two main systems that are often used for scientific research papers: the Vancouver system and the Harvard system.

Remember to

- Check journal requirements — journals require one's citations to be in a particular citation format, and may also require all references to be cited along with the complete title
- Ensure that all citations are accurate, and that no references used have been missed out — check and double-check!
- Try to avoid using Internet resources
- Use original sources as much as possible

Vancouver system

In this system, the references are numbered according to the chronological order in which they are quoted in the text. They should not be ordered alphabetically, but in numerical sequence to ensure sequential numbering in the text. It is vital that the numbers appear in correct sequence throughout the text, e.g., 1, 2, 3 rather than 1, 3, 2 etc.

References present in tables or figures should also be cited, and included in the reference list. References cited in the text should either be numbered within square brackets or as superscript numbers. If the number appears at the end of the sentence, it should appear within the punctuation. It is important to note, however, that once one has decided on a particular style, one sticks to the chosen style throughout the research article. The same reference should also be assigned only one text number throughout the research article. If one wants to use the same source again within the article, one should use the original number assigned to it.

Example: Nather [1, 3] demonstrated that...

It was demonstrated that[1,3]...

Within each reference, the various components such as author name, journal title and date should be recorded in a consistent way.

Harvard system

Here, the author and the date which the paper was written are to be included. The ways in which the authors are to be cited in the paper differ slightly depending on the number of authors to be cited.

If there is only one author to be cited,

- Mention the last name of the author in one's sentence, followed by bracketing the year in which the source was written.

  Example: Nather (2000) demonstrated that...

- Alternatively, one could bracket both the last name of the author and the year in which the source was written at the end of the sentence.

  Example: It was shown that.... (Nather, 2000)

- If the same author and year date is cited from different articles, use a, b and c in the research paper to distinguish the different publications.

    Example: (Nather 2000a,b)

    For two authors,

- Mention both author's names in one's sentence, followed by bracketing the year in which the source was written. The authors should be quoted in alphabetical order, according to the last name of the authors.

    Example: Nather and Wong (2013) showed that…

- Brackets the last names of both authors and the year which the source was written at the end of the sentence.

    Example: It was established that… (Nather & Wong, 2013)

    If there are three or more authors,

- Mention the last name of the first author, followed by '*et al.*'

    Example: Nather *et al.* showed that…

    If the author is unknown,

- The name of the journal and the date it was published should be used. Example: There was clear evidence that… (Lancet, 1998). In this case, the source should be listed under "L" in the references section.

At the end of the paper, a "References" list should be inserted. The list should not be numbered, but should be ordered in alphabetical order according to the last names of the authors.

For references starting with the same surname and initials, single-author works should be listed first, followed by two-author works and finally multiple-author works. Within each reference, components such as authors' surnames, initials, journal articles and titles should be recorded in a consistent manner. For multiple-author and two-author works, names of authors should be listed alphabetically.[9]

Editing, Editing and More Editing!

Having completed writing the research article, one's work is still far from finished. In order to achieve a good research paper, one's paper must go through numerous rounds of editing. In fact, it is commonplace for one's research paper to go into its sixth to eighth draft. The main purpose of editing is to:

- Improve the standard of the paper, by editing to ensure the work is written in the best and clearest way possible.
- Delete unnecessary words in the original manuscript and trim tables, figures and legends. The original manuscript, perhaps 10 to 15 pages, may be shortened to a final acceptable length of not more than 6 to 8 pages through editing.

Editing is often a long and tedious process. Tips to ease the editing process include:

- **Aside from editing one's own paper, the primary author should also ask all co-authors to edit the paper before arriving at the final manuscript. Ask all co-authors to edit the paper in addition to the primary author who has written the complete manuscript. Co-authors may have contributed to writing certain sections of the manuscript.**
- **Use the "put on the shelf" technique**

After writing several drafts of the same research article, one would inevitably be exhausted. The primary author should put aside the finished product for a week and go for a break. After about two weeks, one can then revisit the finished product. The author will be

> "The best advice I can give on this is, once it's done, to put it away until you can read it with new eyes. Finish the short story, print it out, then put it in a drawer and write other things. When you're ready, pick it up and read it, as if you've never read it before. If there are things aren't satisfied with as a reader, go in and fix them as a writer: that's revision."
>
> — Neil Gaiman[10], English novelist

surprised that he now sees the same product with new sight or with fresh lenses. Mistakes previously gone unnoticed may now appear. This eventually leads to a better end product.

Summary

Armed with the guidelines and tips for scientific writing, one can now begin writing the research paper. The writing process will definitely be long and tedious one. However, the benefits reaped from completing the research paper would definitely justify the hard work. All the best and happy writing!

References

1. Temple, E. (2013). 'My pencils outlast their erasers': great writers on the art of revision. Retrieved from http://www.theatlantic.com/entertainment/archive/2013/01/my-pencils-outlast-their-erasers-great-writers-on-the-art-of-revision/267011/
2. Turner, J. (2008). How to write a great research paper. Retrieved from Washington University in the St. Louis Applied Research Lab, http://www.arl.wustl.edu/~jst/talks/writingResearchPapers.pdf
3. Wang, L. (2014). Top ten tips for sleep-inducing scientific writing. Retrieved from http://cen.acs.org/articles/85/i37/Newscripts.html
4. Kamat, P. (2010). How to write an effective research paper. Retrieved from Instytut Chemii Fizycznej, http://ichf.edu.pl/educ/MSD/how_paper.pdf
5. Nather, A. (2002). *Writing the Finished Product — An Article for a Journal, Research Methodology in Orthopaedics and Reconstructive Surgery.* Singapore: World Scientific, p. 715.
6. Beck, K. (1999). *Kent Beck's Guide to Better Smalltalk: A Sorted Collection.* United States of America: Cambridge University Bridge.
7. Berry, C. (2013). 70 brilliant Oscar Wilde quotes. Retrieved from http://quotesnsmiles.com/quotes/70-brilliant-oscar-wilde-quotes/
8. Unknown. (2014). Ralph Waldo Emerson quotes. Retrieved from https://www.goodreads.com/author/quotes/12080.Ralph_Waldo_Emerson

9. John Wiley & Sons. (2013). Wiley-Blackwell author services. Retrieved from https://authorservices.wiley.com/reference_text.asp?site=1
10. Unknown. (2014). Neil Gaiman quotes. Retrieved from http://www.goodreads.com/author/quotes/1221698.Neil_Gaiman

Chapter 12

Uncovering the Review Article

Zest Yi Yen Ang & Aziz Nather

What Makes a Good Review Paper?

A review paper provides a summary of the latest developments of the current state of research in a given topic, with recommendations for future research directions. It is an amalgamation of technical information amassed from previously published transcripts, put together to form a coherent piece. Review articles communicate important messages, contributing to intellectual enrichment and enhancing standards of research.[1]

Writing a review article is challenging — it would usually necessitate reading relevant texts and other related pieces in detail, to eventually present a sound and well-informed judgment of the topic.

Review articles give insights by providing an alternative point of view on previously unknown or not well-understood relations among distinct studies. Reviews therefore attract more journal, textbook and thesis citations than any other types of articles, and substantially contribute to the impact factor of journals. Thus, review articles have earned a highly regarded place in scientific research.

Review articles can be classified either according to their mandate or methodological approach.

Mandate[2]

Review articles can be **invited** or **unsolicited submissions.** Invited review articles are written by invited experts in the relevant field of study. Unsolicited submissions are written by choice, after researchers have chosen to study a particular field of interest.

Methodological Approach

Review articles can also be classified according to their method of approach, and are sorted into two main categories: narrative and systematic.

Narrative review articles

Narrative reviews adopt the traditional approach, and does not include a section describing the methods used in the review. Methods of selecting articles are arbitrarily based on the experience and subjectivity of the author, who is often an expert in the area. There is a substantial weakness in the validity of this form of review, since different research workers may employ different methodologies. The absence of a clear and objective methodology section leads to a number of methodological flaws, which can bias the author's conclusions.[3]

Systematic review articles

A systematic review article is one in which the authors have systematically searched for, appraised and summarised all data available in the articles selected. This is done using a clearly defined methodology, which is reproducible.[4]

Reducing bias

Unlike the unstated methodology of narrative reviews, systematic reviews aim to minimise bias[5] by using explicit and pre-selected criteria to obtain objective information.[6] The methodology used is clearly

Fig. 12.1: Levels of evidence.

documented in the "Methods" section, and is detailed enough to ensure it can be reproduced for verification.

Ranked with providing the strongest evidence and is most free from various biases in medical research, systematic reviews top the hierarchy of evidence (Fig.12.1), and are commonly used as evidence-based medicine.[7]

The key differences between systematic and narrative reviews are summarised in Table 12.1.

Functions of Review Articles

Review articles are written for the following reasons:

  i. to **organise literature** and relate it to your research topic
 ii. to **synthesise results** into a **summary** based on what is and is not known
iii. to **rank** literature based on significance
 iv. to **analyse** the contents of articles

Table 12.1: Narrative *vs.* systematic reviews.

	Narrative	Systematic
Scope of content	Provides a broad overview of an area of research	Investigates a clearly defined clinical question
Transparency of search	No "Methods" section	Includes a "Methods" section
Search criteria	Search protocols or selection criteria for selecting evidence are seldom reported	Studies selected for review using an explicit protocol that specifies inclusion and exclusion criteria
Scope of articles to review	Only refers to published data	Refers to unpublished data to minimise the risk of publication bias
Degree of objectivity	More prone to author's bias	Less prone to author's bias
	Results are analysed on a qualitative level	Results are analysed on a quantitative level

    v. to **evaluate** literature — identify the strengths and weaknesses of a text, based on specific criteria
    vi. to **identify patterns and trends** in literature
    vii. to **identify inconsistencies, contradictions** and **research gaps**, and henceforth **recommend** new research areas
    viii. to **draw a conclusion** based on the information gathered from various resources

Choosing a Topic

One must think carefully about the topic to be chosen for review. It is important to consider the following factors:

- **Have significant clinical impact**
  Your review article should provide evidence that a new treatment is helpful for clinical practice. It could also provide evidence that a standard treatment is no longer helpful.

- **From a well-studied field**
  An area having many more authors, perspectives, theories and controversies is easier to study than one involving only a few people.
- **Be of current interest**
  Pick a timely topic that is currently of keen interest in research.

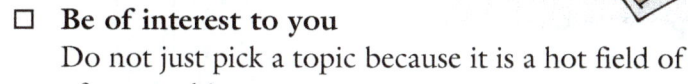

- **Be of interest to you**
  Do not just pick a topic because it is a hot field of study. Pick one of personal interest.
- **Narrow research topic**
  It must not be too broad. Estimate how long of a review you want to produce.
- **Be controversial**
  Make sure you have something to evaluate. You should pick a topic that has at least two competing hypotheses to explain/test it. In this way, one can arrive at a stand.

Literature Review

A sound review article is characterised by a thorough and disciplined literature search. It is important to take into account **all relevant studies,** to provide a holistic view and weed out bias.

Article searches can be performed in the following:

1) **Computerised literature in electronic bases**
   a. Typically, published papers and abstracts are identified by a computerised literature search of electronic databases that can include **PubMed** and **Cochrane Central Register of Controlled Trials (CENTRAL)**. It is recommended to present a full electronic search strategy for at least one major database to be presented.[8]

b. **Reference lists**
Located at the end of every research paper, reference lists provide a list of references cited by the research paper. They are treasure troves of good websites and sources to refer to. By checking the reference list of each research paper, you can easily find other related and useful chapters, articles and web pages to extract valuable information from.

c. **Hardcopy searches**
Look through library resources for relevant papers, books, abstracts and conference proceedings.

d. **Citation index**
One of the best sources there is. It is a compilation of all the articles referenced by recently published articles.

e. **Web searches**
To increase the possibility of securing good articles, it would help to use different databases sources other than Google, such as *Web Crawler*, *Google Scholar* and *Web of Science*. These websites cast a wider net for searches by using several search engines concurrently.

Selecting Articles to Review

How do you Know it is "The One"?

After you have successfully chosen a topic, the next step would be to start the search.

In selecting articles to review, adopt the following methodology:

1) **Start from most recent articles related to the topic**
2) **Select 30**
   Select those with the most important references.

> Do not use material from the Internet unless it is a professional, peer-reviewed scientific journal, of which there are now many on the Internet. Most of these are published by professional associations. If you are not sure of the validity of your source, seek a second opinion.

3) **Read and make a summary for each.**
   Do this in the form of a bulleted list of the conclusions drawn from each figure.
4) **Combine all into a single table.**
   Each research article should occupy a row and the publication issues should be in columns. This allows for easy viewing of which papers agreed on which topics, what trends emerged over time and where the controversies in the field lie.

Afterwards, **rank** each article according to the following criteria (Fig. 12.2):

- **Relevance** to topic
  Is the content of the article suitable for your topic? Does it answer the right questions?
- **Quality** and **accuracy** of article
  Does the information provided have contradictions or conceptual errors which would undermine its credibility?
- **Clarity** of article
  Is it understandable, or does it hide behind of wall of scientific jargon and terminologies?
- **Content**
  Is the article substantial enough?
- **Reliability**
  Is it oft-cited? Is the research from a credible institute?

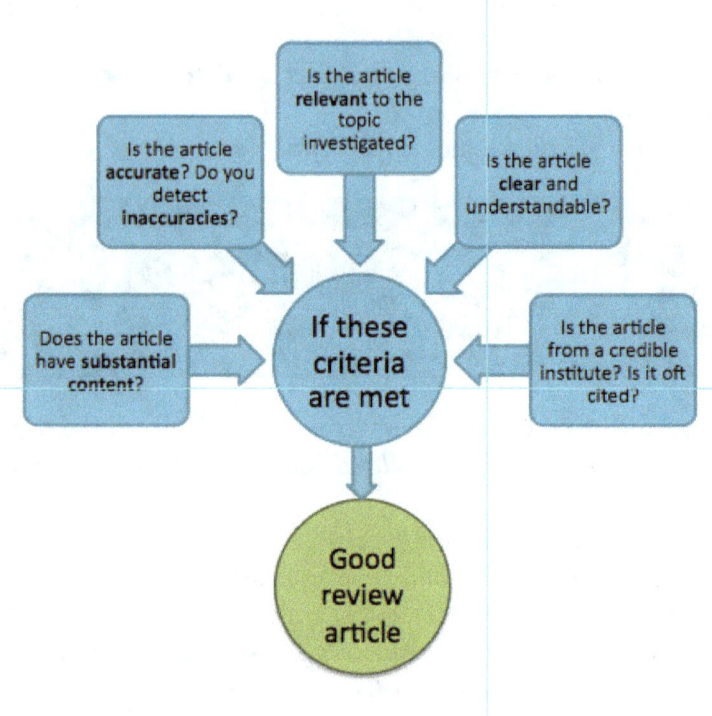

Fig. 12.2: Process in selecting an article to review.

Analysis of the Articles

The next step would be to start dissecting the articles you have selected. Show your understanding of an article's ideas and develop a thoughtful response to ideas that bring up key points of conflict.[9]

To aid your evaluation of the articles, it would help to ask yourself the questions as listed in Fig 12.3.

Writing the Review Article

After analysis and evaluation of the research articles, it would be time to begin writing your article. A review typically consists of a general introduction of the context. Towards the end, the main points covered and take-home messages are reiterated.[10]

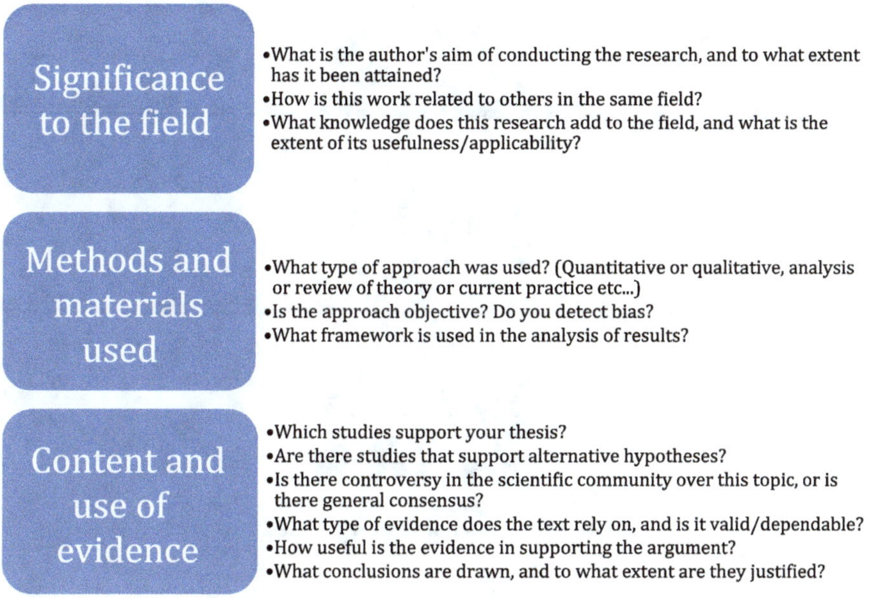

Significance to the field
- What is the author's aim of conducting the research, and to what extent has it been attained?
- How is this work related to others in the same field?
- What knowledge does this research add to the field, and what is the extent of its usefulness/applicability?

Methods and materials used
- What type of approach was used? (Quantitative or qualitative, analysis or review of theory or current practice etc...)
- Is the approach objective? Do you detect bias?
- What framework is used in the analysis of results?

Content and use of evidence
- Which studies support your thesis?
- Are there studies that support alternative hypotheses?
- Is there controversy in the scientific community over this topic, or is there general consensus?
- What type of evidence does the text rely on, and is it valid/dependable?
- How useful is the evidence in supporting the argument?
- What conclusions are drawn, and to what extent are they justified?

Fig. 12.3: Outline of evaluation criteria.

General Structure of a Review Article

The general structure of a review article is as follows[11] (Fig 12.4):

i. **Introduction**
ii. **Body**
   1. → Summary
   2. → Critique
iii. **Conclusion**

Introduction

Length

An introduction is typically one paragraph long for a journal article review and two or three paragraphs long for a longer book review. It should be less than 1/5 the length of the paper.

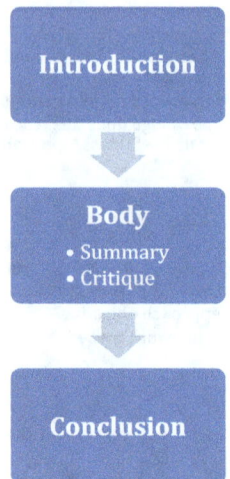

Fig. 12.4: Flowchart of review article structure.

Elements

Introductions should include a few opening sentences that announce the author(s) and the title, and briefly introduce and explain both the topic and thesis. Provide necessary background information required to understand the upcoming discussion. Present the aim of the text and summarise the main finding or key argument, and outline the order in which you will discuss it. Conclude the introduction with a brief statement of your evaluation of the text. This can be a positive or negative evaluation or, as is usually the case, a mixed response.

Body of the paper

The body usually consists of two parts: the summary and the critique.

Summary

This consists of a summary of the main points of the article in the same order the writer uses. Omit any descriptive details such as statistics and examples, and include only the main points. Use clear, incisive sentences that can deliver the message effectively.

You must rephrase and refrain from using the same word combinations that the author has used. If you use quotations from the article, use quotation marks to make this clear. Plagiarism is serious and an egregious mistake to avoid at all costs.

Critique

The critique section can be classified as the most important part of the review article, where you critique and value-add to the primary literature you have sourced for. Give evidence to substantiate your interpretation of the data and the stand you would eventually take. A critical response should be balanced and well considered including both positive and negative statements. It is important to include other referenced sources to support your evaluation, and to link and compare between studies.

Your evaluation should be based on certain criteria, such as the following:

1. What is the main point or argument in the article?
2. Do the points of other writers concur with or differ from the arguments in the article?
3. Do the writer's ideas help or hinder their argument?
4. Are the methods used sufficient to meet the study's aims, or reported in such a manner that the study's conclusions can be relied upon? (e.g., is it a double-blinded and randomised clinical trial?)
5. Does the writer possess any bias?
6. Is the writer qualified to make such claims?

Organise your content

When writing your article, it would help to draw a **conceptual scheme of the review**,[12] e.g., with mind-mapping techniques. Such diagrams can help recognise a logical way to order and link the various sections of a review. A careful selection of diagrams and figures relevant to the reviewed topic can be very helpful to structure the text too.

Sort and arrange your ideas into paragraphs. Arrange your paragraphs such that they convey one idea at a time, for example, one paragraph per methodology evaluated. End your paragraph with a stand regarding the extent to which the point is still applicable.

Use **topic headings** to indicate to the reader what concepts or ideas will be covered in that section. Your headings should be informative, such that anyone reading just the headings in the article will be able to get a brief idea on the structure and organisation of the article.

Include a paragraph on **recommendations** on how the research can be improved, in terms of research methodology, ideas and analytical frameworks used.

Incorporate feedback from peer reviewers

Reviews of the literature are normally peer-reviewed in the same way as research papers. Having read the review with a fresh mind, reviewers may spot inaccuracies, inconsistencies, and ambiguities that had not been noticed by the writers. It is thus critical to incorporate feedback from reviewers in shaping your final manuscript to produce the best possible results.

Conclusion

Your conclusion should leave a lasting impression on the reader, and gain their interest for further developments in that field.

The conclusion is typically a very short paragraph, in which you firmly state your overall stand for your thesis. Succinctly summarise all main points covered, and point out the significance of these results.

Ensure that you have a clear take-home message that integrates the points discussed in the review.

Additional Items to Look Out for

Captions

Captions are one of the most important elements of a review article. A reader can take one look at the captions and decide whether the article is worth the read. Hence, a caption should be clear and succinct, yet detailed, and should convey all the information needed for a reader to understand the figure, without reading the whole manuscript.

Good captions *do not regurgitate information in a figure/table*, but explain *what it means* and *why it is important*.

Captions have a lot of information to relay, so they must be longer than one or two sentences, but should not be longer than about 10 sentences.

Figure 12.5 summarises the main ingredients making up a good review article.

Now that you have been well briefed on how to write a review article, it is time to put your skills to the test. All the best!

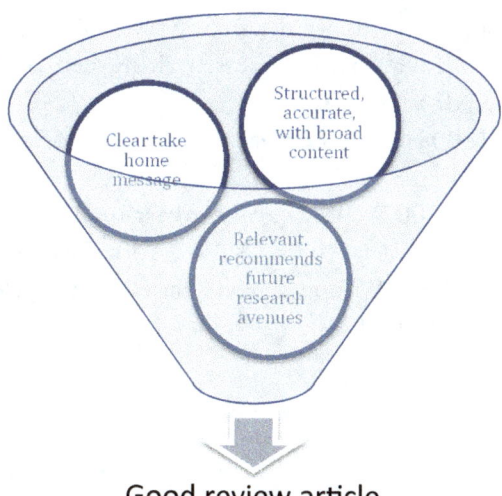

Fig. 12.5: Ingredients of a good review article.

References

1. The PLoS Medicine Editors. (2006). The impact factor game. *PLoS Med.* **3**(6): e291, doi:10.1371/journal.pmed.0030291
2. Noguchi, J. (2006). *The Science Review Article: An Opportune Genre in the Construction of Science.* Bern: P. Lang.
3. Mulrow, C. D. (1987).The medical review article: state of the science. *Ann. Intern. Med.* **106**: 485–488.
4. Law, K. & Howick, J. (2014). OCEBM table of evidence glossary. Retrieved from http://www.cebm.net/index.aspx?o=1116
5. Oxman, A. D. & Guyatt, G. H. (1993). The science of reviewing research. *Ann. N. Y. Acad. Sci.* **703**: 125–133; discussion 133–134.
6. Murphy, C. (2012). Writing an effective review article. *J. Med. Toxicol.* **8**:89–90.
7. Burns, P, B., Rohrich, R. J. & Chung, K. C. (2011). The levels of evidence and their role in evidence-based medicine. *Plast. Reconstr. Surg.* **128**(1): 305–310.
8. Liberati, A., Altman, D. G., Tetzlaff, J., Mulrow, C., Gøtzsche, P. C., Ioannidis, J. P., Clarke, M., Devereaux, P. J., Kleijnen, J. & Moher, D. (2009). The PRISMA statement for reporting systematic reviews and meta-analyses of studies that evaluate health care interventions: explanation and elaboration. *J. Clin. Epidemiol.* **62**(10): e1–e34.
9. University of Fraser Valley (n.d.). Article review/critique. Retrieved from http://www.ufv.ca/media/assets/writing-centre/Article+review+and+critique.pdf
10. Ridley, D. (2008). *The Literature Review: A Step-by-Step Guide for Students.* London: SAGE.
11. Pechenik, J. A. (2007). *Writing Summaries and Critiques. A Short Guide to Writing About Biology* (6th ed.). New York: Pearson, pp. 130–138.
12. Bem, D. J. (1995). Writing a review article for *Psychological Bulletin*. *Psychol. Bull.* **118**(2): 172.

Chapter 13

Writing a Case Report

Zest Yi Yen Ang & Aziz Nather

What is a Case Report?

Case reports represent the oldest and most familiar form of medical communication. Given the unpredictable and challenging nature of medicine, many medical students will have come across a patient who has not been a textbook case. The patient may have had an unusual presentation, strange new pathology or unforeseen reaction to a medical intervention.

Since the time of Hippocrates from 460 BC to 370 BC, medical case histories have been documented to advance the base of knowledge in clinical medicine. Famous case studies have also helped shape the way we view health and disease.[1-3]

In medicine, a case report is a detailed report of the symptoms, signs, diagnosis, treatment and follow-up of an individual patient. Case reports may contain a demographic profile of the patient, but usually describe an unusual or novel occurrence worth knowing. Some case reports also contain a literature review of other reported cases.

Case reports are often accompanied by photos or figures that give readers the chance to make their own decisions about whether the diagnosis was correct.[4]

What is the Purpose of a Case Report?

However, some argue that case reports are increasingly irrelevant in as medicine takes a slant to become more evidence-based. Many medical journals have stopped publishing case reports. The rarity of cases targets only a specialised few, and hence case reports contribute little to everyday medical practice. Their anecdotal nature lacks the scientific rigour of large, well-conducted studies, and they have therefore fallen down the hierarchical ladder of medical evidence (Table 13.1).[5]

Nonetheless, case reports still have a role to play in furthering medical knowledge and education in the following ways:

Quick Form of Communication between Scientists

Case reports are published quickly in comparison to randomised control trials,[6] allowing for rapid and short communication between clinicians. They are also less costly and time-consuming compared to large-scale research.[7]

Provide New Insights

Because remarkable and atypical cases are less likely to be published, case reports permit discovery of new diseases and unexpected effects

Table 13.1: Level of evidence.

Category I	Evidence from at least one properly randomised controlled trial
Category II-1	Evidence from well designed controlled trials without randomisation
Category II-2	Evidence from well-designed cohort or case-control analytic studies, preferably from more than one centre or research group
Category II-3	Evidence from multiple case series with or without intervention or dramatic results in uncontrolled experiments
Category III	Opinions of respected authorities, based on clinical experience, descriptive studies, and **case reports** or reports of expert committees.

(adverse or beneficial) as well as the effectiveness of new surgical or treatment methods. They describe important scientific observations otherwise missed or undetected in clinical trials.

They therefore play an important role in medical education, by bringing to light new diseases, anomalous patient behaviour, unexpected effects of drugs and their method of administration and even different techniques to treating typical conditions.[8] They therefore provide many new ideas in medicine.[9]

Giving an All-Rounded Understanding of an Illness

Randomised clinical trials usually only inspect one variable or very few variables. They rarely reflect the full picture of a complicated medical situation. Conversely, the case report can detail many different aspects of the patient's medical situation (e.g., patient history, physical examination, diagnosis, psychosocial aspects, follow-up).[6] Different areas of medical education such as physiology, pathology, pharmacology and anatomy are brought together in case reports. This helps students and doctors develop a more holistic approach to patients.

Case reports also are useful in other ways (Fig. 13.1):

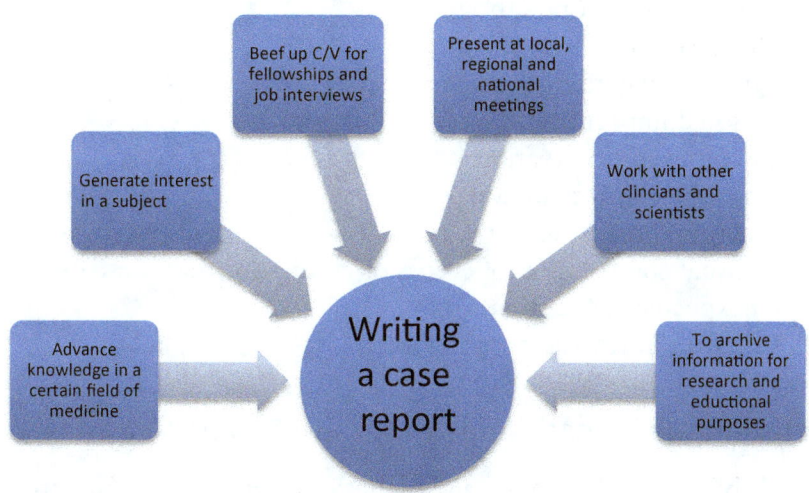

Fig. 13.1: Other benefits of a case report.

Choosing the Topic for a Case Report

Be attentive and constantly look out for novel cases when on ward rounds or during clinic sessions. Should you not know what constitutes a good case report topic, refer to a senior doctor for advice. They may also have cases that you can help research and write on.

Case reports can be on the following content[10]:

a. An unexpected association between diseases or symptoms or between two illnesses thought to be mutually exclusive
b. An unexpected event in the course of observing or treating a patient
c. Findings that shed new light on the possible pathogenesis of a disease or an adverse effect
d. Unique or rare features of a disease
e. Unique diagnostic or therapeutic approaches
f. An unrecognised link between two diseases thought to be mutually exclusive
g. A positional or quantitative variation of the anatomical structures
h. Adverse response to therapies
   - Unusual side effects or interactions
i. Shed new light on pathogenesis
j. Illustration of a new theory
k. Question regarding a current theory
l. Personal impact
m. Uncommon observations
n. Unexpected presentation
o. Emerging disease
p. Unusual combination of events or conditions that cause confusion
q. Unexpected association
r. Variation in the disease process

Gathering Information for your Case Report

To establish a credible and validated case, assemble information from varied and credible sources.

Perform a Literature Search for your Case

Before you begin, perform a literature search on similar cases to collate information to give you varied perspective and set you in the right direction. Review relevant scientific publications: abstracts, original science, review articles as well as relevant reference textbook articles. Medical database such as PubMed, Ovid or Medline can be used to check if there have been any similar cases; this helps you gauge how rare your case is.

To keep yourself organised, make a reference list of the articles to refer to as you write your case presentation. To do so, assemble case information and literature reviews into a file, then assign each article a reference number to form a list. Write a report with references to the indexed articles.

Liaise with Doctors in Charge

Discuss the case report with senior doctors in charge, not just to gain their permission, but also to obtain their guidance and advice.

Approach clinicians of other specialties who are involved in the care of the patient. This allows you to substantiate your case with detailed patient information such as:

- Case history
- Hospital progress notes
- Lab reports
  - X-rays
  - Pathology reports
- Discharge summaries
- Outpatient progress notes

Gain Consent

It is of utmost importance that advancement of science does not occur at the expense of the rights of the patient.[11] Patient confidentiality must be preserved even when writing the case report. Information that identifies a patient should not be published, and informed consent should be obtained if there is any possibility of patient confidentiality being breached.

During the consent process you must explain why you wish to share their case with others, the risks and benefits of doing so, and answer any questions the patient may have. Get the help of a senior if you require it.

Selecting the Right Journal

Always begin with the end in mind. Define your target audience by identifying 3–4 suitable journals, taking into consideration the following:

- Audience
- Field
- Impact factor
- Likelihood for publication

After selecting the right journal, review previous samples of published case reports from the journal for exemplary models to follow.

Always review the "Instructions for Authors" from the journal chosen before beginning to write, which vary among journals. Journals typically specify what content has to be covered, the word limit for each section, the number of tables and figures used and the number of references cited.[1]

For example, the *British Medical Journal* (*BMJ*) has the following criteria[12] (Table 13.2):

Table 13.2: Instructions for publishing in the *BMJ*.

Authorship	Maximum four
Summary	Max 150 words
Images	To be submitted in colour and in the following formats: jpg, tiff, gif, PowerPoint and eps
Competing interests	Compulsory to declare any financial gain or personal rivalry

Writing the Case Report

Do a quick rough first draft of the following sections in the order below:

- ✓ Case
- ✓ References
- ✓ Introduction/background
- ✓ Discussion
- ✓ Abstract (normally easier to write after completing the main body of text as you would be more familiar with the contents of the case)
- ✓ Title

Title

The title should be informative and accurate. It should also be succinct, to effectively convey the essence of the case report.

The title should also include the words 'case report', and 3 to 5 key words. This facilitates retrieval with electronic searching.

Abstract

The abstract is typically required by most journals when submitting a case for publication. The abstract is the "face" of the report — it is indexed by most electronic databases and is what represents your case in searches. Hence, it is important to include key words throughout the abstract, so that your article can be more easily found during searches. Reviewers and judges rate a case report's merit based on the abstract.

Your abstract is a summary of the following:

— The objectives of the report
— Case presentation
   o Diagnosis
   o Intervention (diagnostic, therapeutic or preventive)
   o Outcome
— Understanding and implications

Keep your abstract within the word limit as determined by the relevant journal.

Introduction

As this constitutes the beginning of the essay, the introduction must capture the interest of the reader. Present the relevance of your article to entice busy physicians to take an interest in your report. The introduction should not exceed three paragraphs.

You should include the following in your introduction

- An overview of the medical condition
- A single sentence describing the patient's condition
- Rationale for reporting the case, adequately substantiated
- Justify why the case report is novel and deserving of mention, supported by references.
  - *Previously unreported?*
  - *A new pattern?*
  - *Previously unsuspected relationship?*
  - *Unusual diagnosis, prognosis, therapy, harm?*
- Contextual information required for readers to better understand the text

Case Presentation

Your case presentation should succinctly list out information pertaining to the patient's case — mention only important information, and leave out superfluous data like normal vital signs and other irrelevant patient information to avoid confusing readers.

Case presentations should cover the following[13]:

- Medicine three months prior to the study, to eliminate any possibility of after effects
- All records of the patient's diet, as food can interfere with the pharmacologic effect of the drugs. Always suspect food allergy first before the possibility of drug allergy.
- One or two images to engage your reader

Ensure your case presentation is relevant and thorough enough such that readers without background knowledge will be able to infer a conclusion.

Discussion

The discussion is the most important section in which you value add to what was provided in the patient information.

Reviewers will be interested in the evidence proving your case is rare. Hence, the discussion should centre on why the case is different or unique, such as unprecedented revelations in disease progression, therapy and treatment outcomes. Give the **learning points** of the case, as well as alternative justifications and novel hypotheses about a condition. Review why certain decisions were made and extract lessons.[14]

Evaluate the case's validity, accuracy and distinctiveness with respect to an extensive supply of already published literature. Establish a casual relationship between findings and results from other case studies, and validate its credibility on a scale.

Do include supplementary parts such as tables, figures, graphs and illustrations, which provide essential data and enhance flow and clarity of the article.

Draw recommendations and conclusions on how future clinical practice will be affected based on the features of the case. Be sure to substantiate your case.

Patient's Perspective

The patient should share his or her perspective or experience whenever possible. This provides deeper insight on the symptoms of the illness in question and preempts doctors on the signs to look out for.[15]

Conclusion

The conclusion should be brief and end with a key take-home message. It should also include evidence-based and justified recommendations to clinicians or scientists, and clearly state the future prospects and likely effects on the medical world.

References

Do not be tempted to give a lot of references — remember that more is not always better as there are space limitations. To be sure, check "Instructions for Authors" provided by the journal receiving your case report.

Select your sources wisely by using credible databases like PubMed, Jstar and Medscape. Avoid referencing abstracts.

Authors' Contributions

Authors make substantial and consequential contributions to the case report. Before you begin, you should decide on authorship and determine who you would want to work with.

The first author typically contributes the bulk of the writing, and should be listed first. Subsequent authors should be listed in order of contribution. The senior author, typically the instigator of the case report, should be listed last.

It is compulsory for all co-authors to have given approval on the manuscript prior to submission.

Publishing your Case Report

As case reports rank low on the evidence scale, most journals these days are reluctant to publish case reports.

However, the vastness of cyberspace has allowed for the development of a new breed of medical journals. A number of new online journals such as *BMJ Case Reports*, *Cases Journal*, the *Journal of Medical Case Reports*, *Radiology Case Reports* and the *Journal of Dermatological Case Reports*, allow for the publication and dissemination of notable case reports.[16]

Although still in their infancy, these journals have the potential to act as large case banks that allow doctors to search for cases similar to ones that may be puzzling them, to help guide in their management.

Summary of Approach to Writing a Case Report[17]

1. Identify a patient with an interesting case
2. Do a background literature search
3. Pin point a key take-home message to convey from the case
4. Select the appropriate journal for publishing
5. Obtain instructions on writing from journal
6. Select your authors
7. Perform a second, more specific literature search of specific journals
8. Assemble case information and literature reviews into a file
9. Assign each article a reference number to form a list
10. Write a report with references to the indexed articles
11. Check spelling and grammar
12. Ask a consultant to proof read for you
13. Submit manuscript with cover letter providing personal details (address, phone and fax numbers, e-mail address). Signatures of co-authors are normally required as well.
14. If article is not accepted by journal, obtain reviewer's comments.
15. Perform one or two revisions based on reviewers' comments given
16. Submit article to second journal.[17]

The following is a checklist for a case report write-up:

Checklist
- Due date is ___
- Number of copies needed is ___

1. Abstract
- Introduction and objective
- Case report
- Discussion
- Conclusion

2. Introduction
- Describe the subject matter
- State the purpose of the case report

- Provide background information
- Provide pertinent definitions
- Describe the strategy of the literature review and provide search terms
- Justify the merit of the case report by using the literature review
- Introduce the patient case to the reader
- Make the introduction brief and less than three paragraphs

3. Patient case presentation

- Describe the case in a narrative form
- Provide patient demographics (age, sex, height, weight, race, occupation)
- Avoid patient identifiers (date of birth, initials)
- Describe patient's complaint
- List the patient's present illness
- List the patient's medical history
- List the patient's family history
- List the patient's social history
- List the patient's medication history before admission and throughout the case report
- Ensure that the medications history includes herbals, vaccines, depot injections and non-prescription medications and state that the patient was asked for this history
- List each drug's name, strength, dosage form, route and date of administration
- Verify the patient's medication adherence
- Provide renal and hepatic organ function data in order to determine the appropriateness of medication dosing regimens
- List the patient's drug allergy status, including the name of the drug (brand or generic) and date and type of reaction
- List the patient's adverse drug reaction history and the dates of reaction
- Provide pertinent serum drug levels and include the time of each level taken and its relationship to a dose
- Provide the patient's dietary history

- Provide pertinent findings on physical examination
- Provide pertinent laboratory values that support the case
- Provide the reference range for laboratory values that are not widely known or established
- List the completed diagnostic procedures that are pertinent and support the case
- Paraphrase the salient results of the diagnostic procedures
- Provide photographs of histopathology, roentgenograms, electrocardiograms, skin manifestations or anatomy as they relate to the case
- Obtain permission from the patient to use the patient's photographs or follow institutional guidelines
- Provide the patient's event in chronological order
- Ensure a temporal relationship
- Ensure a causal relationship
- Ensure the patient case presentation provides enough detail for the reader to establish the case's validity

4. Discussion

- Compare and contrast the nuances of the case report with the literature review
- Explain or justify the similarities and differences between the case report and the literature
- List the limitations of the case report and describe their relevance
- Confirm the accuracy of the descriptive patient case report
- Ensure a temporal relationship
- Ensure a causal relationship
- Report the validity of the case report by applying a probability scale such as the Naranjo nomogram
- Summarise the salient features of the case report
- Justify the uniqueness of the case
- Draw recommendations and conclusions

5. Conclusion

- Provide a justified conclusion
- Provide evidence-based recommendations

- Describe how the information learned applies to one's own practice
- List opportunities for research
- Ensure that this section is brief and does not exceed one paragraph

References

1. Kyle, R. A. (2000). Multiple myeloma: an odyssey of discovery. *Br. J. Haematol.* **111**: 1035–1044.
2. Heller, S. (2005). *Freud A to Z.* New Jersey: Wiley.
3. Schiller, F. (1992). *Paul Broca: Founder of French Anthropology, Explorer of the Brain.* Oxford: Oxford University Press.
4. Graf, J. (2008). Handbook of biomedical research writing: the clinical case report. Retrieved from http://www.hanyangowl.org/media/biomedical/clinicalcasereport.pdf
5. Centre for Evidence Based Medicine. (2014). Homepage. Retrieved from http://www.cebm.net
6. Yitschaky, O., Yitschaky, M. & Zadik, Y. (2011). Case report on trial: do you, Doctor, swear to tell the truth, the whole truth and nothing but the truth? *J. Med. Case Rep.* **5**(1): 179.
7. The Evidence-Based Medicine Working Group. (2002). *User's Guide to the Medical Literature: A Manual for Evidence-Based Practice.* Chicago, III: American Medical Association Press.
8. Mason, R. A. (2001). The case report — An endangered species? *Anaesthesia* **56**: 99–102.
9. Harris, R. P., Helfand, M., Woolf, S. H., Lohr, K. N., Mulrow, C. D., Teutsch, S. M., *et al.* (2001). Current methods of the US preventive services task force: a review of the process. *Am. J. Prev. Med.* **20**: 21–35.
10. Cohen, H. (2006). How to write a patient case report. *Am. J. Health Syst. Pharm.* **63**: 1888–1892.
11. Levine, S. B. & Stagno, J. S. (2001). Informed consent for case reports: the ethical dilemma of right to privacy versus pedagogical freedom. *J. Psychother. Pract. Res.* **10**: 193–201.

12. BMJ Case Reports — Instructions for authors. (n.d.). Retrieved from http://casereports.bmj.com/site/about/guidelines.xhtml
13. Gagnier, J. J., Kienle, G., Altman, D. G., Moher, D., Sox, H., Riley, D. & the CARE Group. (2013). The CARE guidelines: consensus-based clinical case reporting guideline development. *J. Med. Case Rep.* **7**: 223, doi: 10.1186/1752-1947-7-223
14. Chelvarajah, R. & Bycroft, J. (2004). Writing and publishing case reports: the road to success. *Acta Neurochir.* (Wien) **146**(3): 313–316.
15. Saleem, M. A. & Macdonald, R. L. (2013). Cerebral aneurysm presenting with aseptic meningitis. *J. Med. Case Rep.* **7**: 244).
16. Anwar, R., Kabir, H., Botchu, R., Khan, S. A. & Gogi, N. (2004). How to write a case report. *Student BMJ* **12**: 45–88.
17. Brodell, R. T. (2000). Do more than discuss that unusual case. Write it up! *Postgrad. Med.* **108**(2): 19–20.

Chapter 14

Writing a Thesis or Dissertation

Zest Yi Yen Ang & Aziz Nather

What is a Thesis or Dissertation?

A **thesis** or **dissertation** is a document submitted in support of candidature for an academic degree or professional qualification. It presents the author's research and findings.[1]

The Significance of a Thesis and a Dissertation

Writing a thesis or dissertation is perhaps the most daunting part of graduate education.

Writing a thesis or dissertation prepares the student to be a professional in the discipline. Through this process, one learns and demonstrates the ability to conduct independent, original, and significant research. The dissertation thus shows that the student is able to[2]:

— **Identify** problems
— **Generate** questions and hypotheses
— **Review** and **summarise** the literature
— **Apply** appropriate methods
— **Collect** data properly
— **Analyse** and **judge** evidence

— **Discuss** findings
— **Produce** publishable results
— **Engage** in a sustained piece of research or argument
— **Think** and **write** critically and coherently

A thesis or dissertation thus marks the culmination of thousands of hours of training, research and writing, and represents you for years after graduation.

Preparing your Thesis

Brainstorm for Ideas

Most research begins with a question. Think about the topics and theories you have studied in your programme, the areas you are interested in and what you would like to know more about. Are there some questions you feel the body of knowledge in your field does not answer adequately?

To come up with questions, you have to know the area well. Thus, read voraciously in the areas that interest you.[3] As you read, look out for the following:

- *Are there interesting contrasts or comparisons or patterns emerging in the information?*
- *Is there something about the topic that surprises you?*
- *Do you encounter ideas that make you wonder why?*
- *Does something an "expert" says make you agree or disagree?*

Set up appointments to talk with senior faculty in your field of significance who can better advise you and help you start narrowing down on potential topics.

You may also refer to archives of theses for examples, to get a better sense of the scope and the traits of a good thesis. Sometimes, older studies can be re-examined in a new context or with more current methods, and can be a new source of topics.[4]

Writing your Thesis[5, 6]

Order of Writing

Your thesis is not written in the same order as it is presented in. It is recommended for you to write your thesis in the following order to ease the writing process:

1. **Preliminary version of background:** Before you begin, think of your objective and motivation behind carrying out the research. This will serve as the basis for the introduction in your final paper.
2. **Methods section:** When you begin your research, write up the methods section as you collect your data.
3. **Plots and tables:** Start organising your data into plots and tables. These will help you to visualise the data and to see gaps in your data collection.
4. **Figure captions:** Once you have a complete set of plots and statistical tests, write figure captions for the plots and tables. The captions should stand alone in explaining the plots and tables.
5. **Results:** Describe your results, but do not interpret them. Interpretation should be left to the discussion section.
6. **Discussion:** In writing the discussion session, be sure to adequately discuss the work of other authors who collected data on similar scientific questions. Discuss the relevance and flaw in methodology of other authors.
7. **Conclusion:** Build the ideas that were mentioned in the discussion section and try to come to some closure. If some hypothesis can be ruled out as a result of your work, say so. If more work is needed for a definitive answer, say that.
8. **Recommendation:** Make recommendations for further research or policy actions in this section. If you can make predictions about what will be found if X is true, then do so. You will get credit from later researchers for this.
9. **Introduction:** After you have finished the recommendation section, look back at your original introduction. Your introduction

should set the stage for the conclusions of the paper by laying out the ideas that you will test in the paper.
10. **Abstract:** Write your abstract last when you have finalised the direction of your thesis.

Components of a Thesis

Title page

When selecting your title, aim to include

1) A witty catchy phrase
2) Keywords that precisely capture the essence of your thesis, and would show up in Google searches[7]
3) State your **contribution, approach** and **result** in no more than eight words

Abstract

An abstract should not exceed 300 words or one page in length. Print it in single spacing and indicate the author and title of the dissertation in the form of a heading.

It does not normally contain references. When references are necessary, its details should be included in the text of the abstract.

The abstract is preferably written towards the end, when you are familiar with your thesis and thus capable of distilling it. It should be concise and include the following:

- *A description of the problem(s) addressed*
- *Your method of solving it/them*
- *Your results*
- *Your conclusion*

Table of contents

Included typically near the start of the thesis, a table of contents includes the chapter titles, including the subheadings within each chapter.[7] Also add in page numbers to make it easier to find sections quickly.

List of figures and tables

All tables, photographs, diagrams, etc., should be listed in the order in which they appear in the text.

Introduction and thesis statement

The introduction lays the groundwork for the rest of the thesis or dissertation. It outlines the broad field of study and then narrows into the crux of the research problem.

Your introduction should have the following elements:

1) **Context:** Provide background information to allow the reader to understand the context and significance of the issue you are addressing.
2) **Motivation/goal:** The introduction should clearly state the motivation behind the thesis. Include a statement highlighting the pertinence of a scientific problem that your paper solves or addresses. This draws readers in and makes them interested in the rest of your paper.
3) **Research direction:** Give a brief overview of the research undertaken, and inform the reader of the direction of the thesis.
4) **Significance:** Explain the need for research, as well as the significance to the field of interest.[8]
5) **Scope:** Explain the scope of your work.
6) **Your contributions:** As a thesis, there is a larger emphasis on original work and interpretation rather than regurgitation. Your introduction should delineate where your contribution comes in.
7) **Length:** The introduction should only be a page or two long.[8] To aid understanding, break the introduction up into logical segments using sub-headings.

A good introduction may need to go through repeated revisions. It would help to ask someone who is not a specialist to read through and give feedback. Eventually, your introduction should be substantial, yet easy to digest and logical.

Methods

The purpose of the methodology chapter is to give an experienced investigator enough information to replicate the study.

This section should include the following[7]:

Research objective and procedures

You should clearly state the method of investigation you have adopted in the study. Substantiate why you think these methods are appropriate given your research scope and circumstances.

Consider the **benefits** of your chosen method and identify any **disadvantages** and how you overcame them. Discuss any variations in methodology from other similar studies uncovered during the literature review, and conclude with a reflection on the aptness of your methodology.

Participant information

Include details of subjects and sources of data e.g. location, sample size, etc. as well as the description of each instrument used to collect data and details of pilot studies (how and when the studies were carried out).

Results

Present your research findings in this section. Give tables and figures of results to illustrate patterns in the data presented in this chapter. Organise your data and break it down into sections to make the information easier to digest.

Limit your content to the presentation and analysis of collected data, without drawing conclusions or comparing results to other researchers. This would be reserved for the discussion section.

Discussion

The discussion links research findings to literature presented in the literature review. Raise different voices of interest in the research question and explore the findings in light of the literature and different perspectives within it.

You may address the following questions:

- *Are your findings consistent with current theories?*
- *Do they give new perspectives? Or do they contradict?*

Address the literature examined in the literature review in the context of your study, as well as the implications of research findings in educational policies or practice. You may raise questions and give recommendations for further research.

Conclusion

The conclusion extends beyond a summary of the chapters or data you have presented in the main thesis or dissertation. Do not merely repeat information in your thesis — provide a synthesis of the key findings and argument projected by the research. Clearly state your stand on your thesis statement.[9]

Do not include any new information or material, and avoid making this section too long.

Appendices

Any material that should be in the thesis but would break its flow or bore the reader unbearably should be included as an appendix. These items range from interview transcripts, questionnaires and data files.

The list of appendices is typically appended near the end of the thesis, on a single sheet. It is listed in alphabetical order.

Bibliography

A list containing references to all books, articles, journals, web sites, etc. referred to in the research article. The citations should conform to guidelines (e.g., APA guidelines) specified.

Submitting Your Thesis

Doing your Checks

Before submitting your thesis, perform the following checks to ensure your thesis is of a commendable standard.

Proofread your thesis a few times and look out for the following:

1. **Spelling:** Spellcheckers are useful for initial checking, but do not catch homonyms (e.g., hear/here), so you need to do the final check by eye.
2. **Sentence structure:** Make sure that you use complete sentences.
3. **Grammar:** Punctuation, sentence structure, subject–verb agreement (plural or singular), tense consistency, etc.

Let other read and comment on your thesis, and treat their feedback seriously.

Final Submission

- Make three final copies: one to your research mentor and two to your department
- Bind your final thesis
- Double-spacing using 12-point font
- One-inch margins
- Include page numbers
- Print cleanly on white paper
- Print double-sided to save paper

References

1. Lovitts, B. E. & Wert, E. L. (2009). *Developing Quality Dissertations in the Social Sciences: A Graduate Student's Guide to Achieving Excellence.* Sterling: Stylus.
2. Yale Graduate School (n.d.). Writing a thesis or dissertation. Retrieved from http://www.yale.edu/graduateschool/writing/forms/Writing%20Theses%20and%20Dissertations.pdf
3. University of North Carolina (n.d.). Honors theses — The Writing Center. Retrieved from http://writingcenter.unc.edu/handouts/honors-theses/
4. Claremont Graduate University (2014). Choosing a topic for a thesis. Retrieved from http://www.cgu.edu/pages/891.asp
5. Kastens, K., Pfirman, S., Stute, M., Hann, B., Abbott, D. & Scholz, C. (n.d.). How to write a thesis. Retrieved from http://www.ldeo.columbia.edu/~martins/sen_sem/thesis_org.html#Copy
6. Narasimhan, P. & Carnegie Mellon (n.d.). How to write a good PhD dissertation. [Powerpoint slides] Retrieved from http://www.cs.cmu.edu/~priya/ICSOC-PhDSymp-2006-dist.pdf
7. Johnston, K. & Trinity College Dublin. (2011). Dissertation writing — a practical guide. Retrieved April 15, 2014, from http://www.tcd.ie/Education/courses/masters/MEdDissertationWriting1112[1].pdf
8. Childers, L. (2008). Guidelines for writing a thesis or dissertation. Retrieved from http://www.jou.ufl.edu/grad/forms/Guidelines-for-writing-thesis-or-dissertation.pdf
9. Assan, J. (2006). Writing the conclusion chapter: the good, the bad, the missing. Retrieved from http://www.devstud.org.uk/downloads/4be165997d2ae_Writing_the_Conclusion_Chapter,_the_Good,_the_Bad_and_the_Missing,_Joe_Assan%5B1%5D.pdf

Chapter 15

What Is Plagiarism?

Eda Qiao Yan Lim & Aziz Nather

Definition

"Plagiarism is the use of others' published and unpublished ideas or words without permission, and presenting them as new and original rather than derived from an existing source."

— World Association of Medical Editors

What is Plagiarism?

Plagiarism is a recurring problem that happens across various fields of work. While many hold the view that plagiarism is an intended act of dishonesty, it can also occur unintentionally due to reasons such as improper citation of publication sources and failure to present borrowed text in the correct format.

The advent of technology has brought about increased accessibility to different forms and sources of publication, further increasing the susceptibility of individuals falling prey to plagiarism.[1] Thus, even though it has long been a concern that plagiarism poses and will continue posing a threat to the originality and progress of scientific research, it has become increasingly difficult to detect and stop such acts.

There are three main forms of plagiarism:

- **Intentionally claiming** works done by others as one's own product
- Using information from a publication **without citation**
- **Failure** to **carry out citation accurately** to acknowledge the publication source[2]

Why is Plagiarism Wrong?

Plagiarism is unethical. Although it may seem like an insignificant act of simply using some information, plagiarism is undeniably an act that goes against morality.[3] This is so as to claim credit or lift ideas from previous publications without proper citation is akin to stealing things that one does not own, which is dishonest.[4] Hence, plagiarism is also otherwise known as literary theft. In addition, plagiarism goes beyond just being a moral issue. Preventing plagiarism is also about learning to see things from different perspectives. When an individual plagiarises, he or she is denying the source publication any credit for the original ideas presented. Would you fancy someone presenting your idea as his or her own without giving you due credit? I believe not. Thus, it is important for the writer to make a conscious effort to carry out proper citation so as to give credit where it is due as a form of respect to the work done by others.[5]

Consequences of Plagiarism

Plagiarism is often the end product of a moment of folly. When committing acts of plagiarism, many individuals fail to realise the possible repercussions that may ensue. These individuals often consider only the short-term benefit of convenience which unfortunately leads to long-term regret. In this light, it is wise to think ahead and recognise that plagiarism serves only to harm in the long run. After all, is it truly worth risking all the time and effort put into completing a research for temporal convenience?

Some possible consequences of plagiarism are:

Destroyed Academic Reputation[6]
Past case study

Jonah Richard Lehrer was a renowned scientific journalist who wrote for the *New York Times*, the *New Yorker* and *Wired Magazine* until it was found that he had self-plagiarised by recycling content from previous publications in several blog posts that he submitted to the *New Yorker* as well as in his book, *Imagine: How Creativity Works*. As a result, he lost his credibility where confidence that the public had in his work plunged significantly and rapidly.[7]

Legal Repercussions

The legal implications of plagiarism can be extremely serious as copyright laws should not be broken. Even though such a result is the least likely, it affects the plagiarist most significantly and negatively. For instance, depending on the severity and extent of plagiarism, it can lead to fines, community service and even imprisonment.[8]

Purpose of Scientific Research Undermined

The very essence of scientific research that encourages critical thinking and originality for the progress of science is also lost with plagiarism. Instead of studying results obtained from experiments carried out to infer possible explanations that may answer the thesis, individuals who plagiarise are unable to tap on their capacity for scientific inquiry. This is evident in our world today, especially so in China. As the *Economist* states, "...China led the world in retractions due to duplication — the same papers being published in multiple journals."[9]

As a result, even though China has contributed an increasing share of scientific publications over the years, she has not been able to improve its citation frequency ranking above the 20th position. This has instead "hindered her scientific ambition" as described by an article published by the National Public Radio, a non-profit membership media organisation.[10] If the problem of plagiarism is allowed to deteriorate, this may well threaten the field of scientific research in the near future, putting scientific progress to a halt.

Types of Plagiarism

These are two types of plagiarism: those with uncited sources (Table 15.1) and those with cited sources (Table 15.2).

Table 15.1:   Uncited sources

1	"The Ghost Writer"	The writer turns in another's work, word-for-word, as his or her own.
2	"The Photocopy"	The writer copies large chunks from the original author without making any changes.
3	"The Potluck Paper"	The writer obtains material from several different sources, with no changes made to the content or phrasing.
4	"The Poor Disguise"	Modifying only certain key words or phrasing of an entire source but copying the entire content.
5	"The Labour of Laziness"	Putting together modified pieces of work from several different sources without coming up with an original piece of work
6	"The Self-Stealer"	Copying from one's own previous works without citing one's own work.

Adapted from: *Turnitin.com* and *Research Resources*[2]

Table 15.2:   Cited sources

1	"The Forgotten Footnote"	Only the name of the author is cited without any other information on location of the source used stated.
2	"The Misinformer"	Wrong information is given about the source.

(Continued)

	Table 15.2:	(Continued)
3	"The Too-Perfect"	"Forgetting to enclose paragraphs which were copied wholesale in quotation marks even though citation has been done correctly".
4	"The Resourceful Citer"	Majority of the paper is done by putting together bits and pieces of information from other research papers even though the sources are cited, without almost none or little original work thought up by the author.
5	"The Perfect Crime"	Citing only certain sources, but paraphrasing material from other research papers without giving the original author due credit.

Adapted from: *Turnitin.com* and *Research Resource*[2]

Guidelines to Avoiding Plagiarism

1. Keep track of your research by accurately recording all information about any publication source used to allow for easy citation.
2. When information is copied wholesale, be sure to place them in quotation marks!
3. Cite the sources of any material used.
4. When in doubt, give credit anyway!
5. Familiarise yourself with your area of research.

Citing Correctly

In-Text Citation

In-text citation refers to citation carried out after the use of any information from an external source. It is important to be specific when it comes to acknowledging the use of any information from a publication as this allows for readers to trace the development and credibility of ideas should the need arise.[5]

Guidelines when carrying out in-text citation

1. After presenting information retrieved from an external source, number the sentences according to the order in which they appear in the text. These numbers will correspond to the numbered

references in the references section (see the next sub-section for more details) at the end of your research paper.

E.g., In 2009, 6.1 million tonnes of waste was generated in Singapore, a 31% increase from the 4.6 million tonnes in 2000.

2. For every sentence containing information from a different source, number the sentence.
3. If different pages contain information from the same source, be sure to also number the separate sentences with the same number.

E.g., (Page 1) In 2009, 6.1 million tonnes of waste was generated in Singapore, a 31% increase from the 4.6 million tonnes in 2000. (Page 2) Due to limited land space of only 699 square km and the lack of natural resources, this increasing amount of waste generated each year is fast becoming a problem for Singapore.

References

Step 1: Identify the type of material (Fig. 15.1)

Step 2: Note the required information for proper citation. The proper details to be cited are listed for printed resources (Table 15.3), electronic resources (Table 15.4) and other resources (Table 15.5).

Step 3: Present the information in the correct format

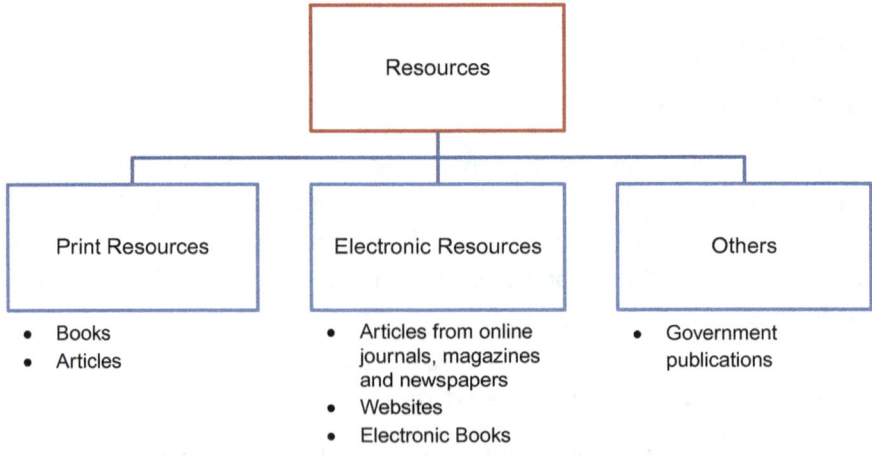

Fig. 15.1: Type of material

Table 15.3: Print resources

Books	Articles
1. Name of author	1. Name of author
2. Publication year	2. Publication date
3. Title of chapter	3. Title of article
4. Title of book	4. Title of publication
5. Place of publication	5. Volume and issue number of publication
6. Publisher	6. Pages on which article is located on

Table 15.4: Electronic resources

Articles from online journals, magazines and newspapers	Websites	Electronic books
1. Name of author	1. Name of author (if available)	1. Name of author
2. Publication date	2. Publication date or date of last update	2. Publication date
3. Title of article	3. Title of website	3. Title of chapter within book
4. Title of publication	4. Date on which information was accessed	4. Title of book
5. Volume and issue number of publication (if available)	5. Website address	5. Place of publication
6. Pages on which article is located on		6. Publisher
7. Digital object identifier (DOI) number		7. Website address
8. Website address (if DOI is not available)		

Table 15.5: Other resources[11]

Government publications
1. Name of government agency
2. Publication year
3. Title of document
4. Place of publication
5. Publisher

Printed Resources

Present the information required in Step 2 in the correct format for 'Books' (Fig. 15.2), and 'Articles' (Fig. 15.3).

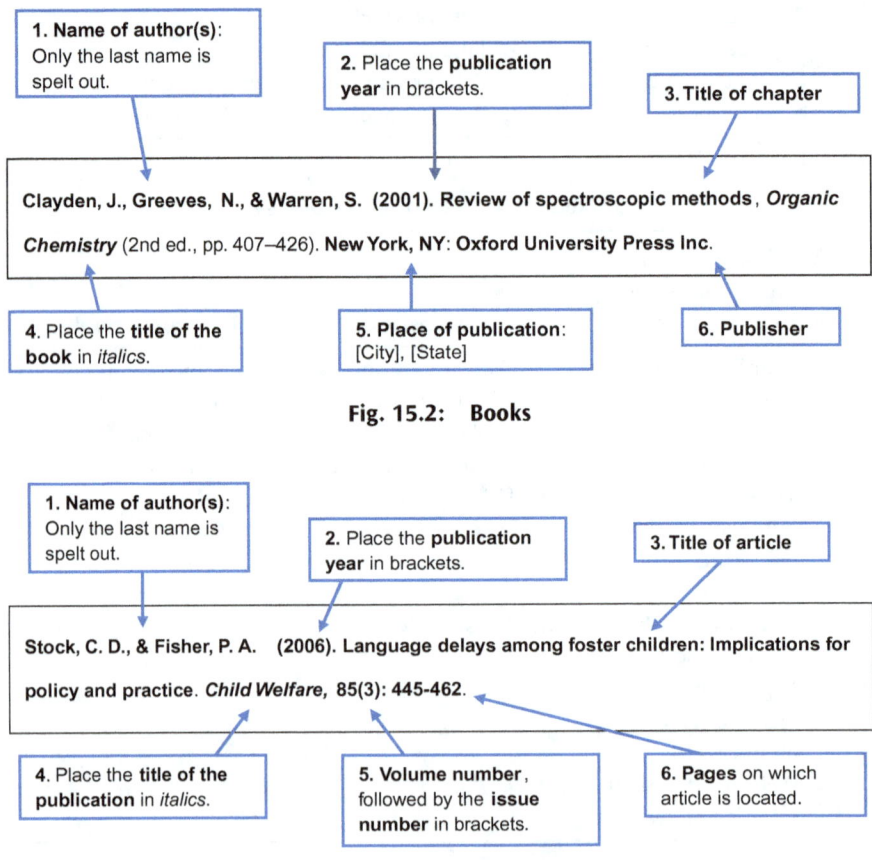

Fig. 15.2: Books

Fig. 15.3: Articles[11]

Example adapted from: http://www.libraries.psu.edu/psul/lls/students/apa_citation.html

Electronic Resources

Articles from online journals, magazines and newspapers

The information must also be cited in the correct format articles from electronic resources (Fig. 15.4) and from websites (Fig. 15.5).

> Overbay, A., Patterson, A. S., & Grable, L. (2009). On the outs: Learning styles, resistance to change, and teacher retention. *Contemporary Issues in Technology and Teacher Education*, 9(3). **Retrieved from http://www.citejournal.org/vol9/iss3/currentpractice/article1.cfm**

Fig. 15.4: Electronic resources

Example adapted from: http://www.libraries.psu.edu/psul/lls/students/apa_citation.html

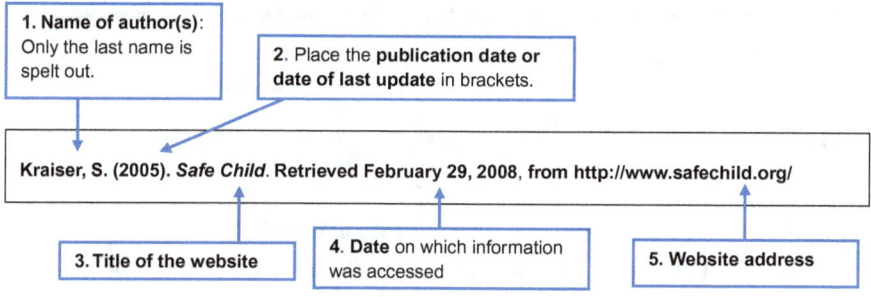

Fig. 15.5: Websites

For websites without authors, simply shift the **title of the website** in *italics* to the front before the publication date or date of last update for proper citation.

Electronic books[11]

The citation format is similar to the one given in the 'Books' section above, except that the address of the website from which the article was retrieved from needs to be included (Fig. 15.6).

> McKernan, B. (2005). *Digital cinema: The revolution in cinematography, postproduction, and distribution*. New York, NY: Mc-Graw Hill. **Retrieved from www.netlibrary.com**

Fig. 15.6: Electronic books

Example adapted from: http://www.libraries.psu.edu/psul/lls/students/apa_citation.html

Other Resources

For government publications, the format adopted is shown in Fig. 15.7.

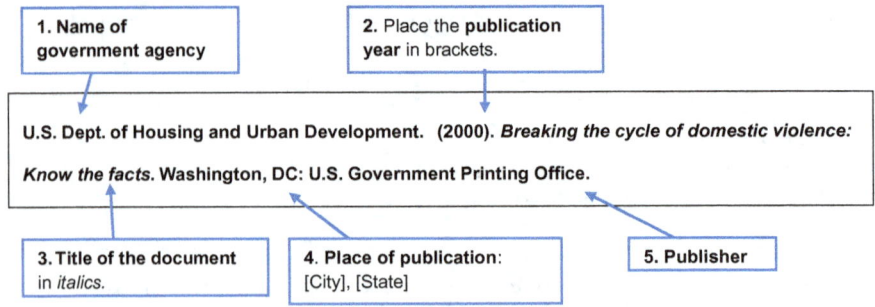

Fig. 15.7: Government publications
Example adapted from: http://www.libraries.psu.edu/psul/lls/students/apa_citation.html

Note: Citations with more than a single line should have a hanging indent of half an inch.

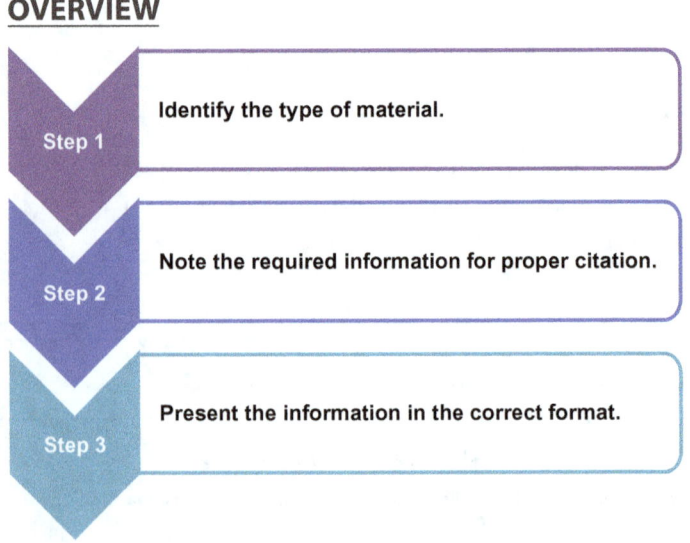

References

1. Masic, I. (2012). Plagiarism in scientific publishing. *Acta Inform. Med.* 20(4): 208–213.
2. Turnitin.com and Research Resources. (2005). What is plagiarism? Retrieved from http://www.mdc.edu/kendall/english/docs/What%20 is%20Plagiarism%20-%20Package.pdf
3. Gotterbarn, D., Miller, K. & Impagliazzo, J. (2006). Plagiarism and scholarly publications: an ethical analysis. *36th ASEE/IEEE Frontiers in Education Conference.* Retrieved from http://blog.lib.umn.edu/swiss/archive/Gotterbarn%20Miller%20%26%20Impagliazzo.pdf
4. Kokemuller, N. (2014). Why is plagiarism unethical? Retrieved from http://everydaylife.globalpost.com/plagiarism-unethical-7240.html
5. Hunter, J. (n.d.). The importance of citation. Retrieved from http://web.grinnell.edu/Dean/Tutorial/EUS/IC.pdf
6. Unknown (2014). Six consequences of plagiarism. Retrieved January 23, 2014, from http://www.ithenticate.com/resources/6-consequences-of-plagiarism
7. Unknown. (2014). Top plagiarism scandals of 2012. Retrieved from http://www.ithenticate.com/plagiarism-detection-blog/bid/89793/Top-Plagiarism-Scandals-of-2012
8. Unknown. (2013). What are the legal consequences of plagiarism. Retrieved from http://copyright.laws.com/copyright-laws/legal-consequences-ofplagiarism
9. The *Economist.* (2013). Scientific research: Looks good on paper. Retrieved from http://www.economist.com/news/china/21586845-flawed-system-judging-research-leading-academic-fraud-looks-good-paper
10. Lim, L. (2011). Plagiarism plague hinders China's scientific ambition. *National Public Radio.* Retrieved from http://www.npr.org/2011/08/03/138937778/plagiarism-plague-hinders-chinas-scientificambition
11. Penn State University (University Libraries). (2013). APA quick citation guide. Retrieved from http://www.libraries.psu.edu/psul/lls/students/apa_citation.html

Section V
Evaluating Your Research

Chapter 16

Reviewing an Article

Aziz Nather

Editorial Process

The Editorial Committee of a Journal comprises the Editor-in Chief, Co-Editors and Assistant Editors. Its Editorial Board is represented by members from different countries in the world. Each member appointed to the Board is a prominent surgeon who has himself contributed significantly to the literature and is well recognised by his peers not only in his country but internationally. The academic standing of the Editors and members of the Editorial Board is crucial to the international recognition that the Journal would receive.

The article received is first checked by the Editorial Office to ensure that the contributor has followed all the instructions required for preparation of the manuscript.

The checklist includes:

- Covering letter to the Editor
- Specified number (usually three) of hardcopies of manuscript, each complete with illustrations
- One floppy disk copy of the typed manuscript in the PC format usually Word Perfect

- Letter of Transmittal duly signed by all authors. This is important to ensure that the authors convey all copyright ownership of this article to the Journal.

The author(s) must affirm that the article is original, is not under consideration by another journal and also that it has not been previously published. This assignment is to take effect only if the work is accepted for publication in the Journal.

In the event that the author has not fulfilled all requirements in this checklist, the article will be returned to the author for re-submission.

It is only when all the requirements have been met that the article is then forwarded to the Editorial Committee for consideration. A letter of acknowledgement of receipt of the article would then be sent to the author, stating that the paper has been received and is under review for possible publication in the Journal.

Each article forwarded to the Editorial Committee is first classified as one of the following:

- Original/Scientific Articles
- Review Articles
- Editorial/Invited Articles/Guest Features

The majority of contributions received are original or scientific papers. A few are review articles. The articles are then further sub-classified into:

- Clinical Contributions
    — Spine
    — Hip
    — Knee
    — Sports Medicine
    — Tumours
    — Paediatric
    — Microvascular Reconstruction

- Experimental Studies
  — Animal Experimentation
    - Histological
    - Biomechanical
  — Cadaveric
  — Tissue Engineering or Cell Culture
  — Immunology

The Editorial Committee after classifying each article would then select the two most appropriate reviewers from the Editorial Board who have experience or interest in the field of study investigated by the author(s) of the article.

The Editorial Office then forwards the article to the two reviewers selected. The reviewer is given the option whether to accept the paper for review or to turn it down. Once the reviewer accepts to assess the article, he is usually given the responsibility to complete the task no longer than usually one to two months upon receipt of the full manuscript.

Reminders and datelines are sometimes sent to reviewers who occasionally fail to meet the deadline.

Assessment by the Reviewer

Each Journal has its own set of criteria for assessment which the reviewer has to use. These include general criteria and scientific criteria. A set format is given to the reviewer performing the evaluation. Detailed comments are required where necessary.

General Criteria

These include:

Originality

A contribution that is new is always welcome and well rewarded by any reviewer. However, even if the work is not original, the article may still be accepted for publication if it conveys worthwhile information on a well-researched topic.

Presentation

A paper written clearly and systematically with a good command of English is always a joy to read compared to one poorly written, lacking organisation or even worse, written in poor English.

Length of article

An average length of an article for an original or scientific paper should be between 4 to 8 pages. Anything more would be too long or too detailed. The article will usually have to be revised to a more concise one. Less than three pages is usually too short for a scientific paper. The reviewer may require more details to be elaborated.

For a case-report, 2 to 4 pages is usually about the right length.

For a review article, 8 to 14 pages is usually requested for a detailed review of the topic assigned.

Illustrations

Good illustrations usually add much to the quality of the article. Photographs of poor quality should be rejected. Well-designed tables would also add to the merit of the article. However, too many illustrations may make the article unnecessarily long. In this case, the editor would select only those illustrations that are absolutely necessary.

On the other hand, there would be articles where the required illustrations are missing. In this case, the assessor will request that such illustrations be included in the revised manuscript.

Bibliography

A critical number of references must be used to reflect an adequate review of literature for the topic studied. The quality of these references is also important. The reviewer may direct the author to look at key and recent articles relevant to the work at hand that he has not assessed.

On the average, most journals will value a large number of references used. However, there are the occasional journals who have returned the manuscript to the contributor requesting for the author to reduce the list of references he has used from say, 37 to only the best 10 to 12. The reviewer will also assess whether the author has referenced the article correctly in the format required. On some occasions if he is unable to assess a particular reference, he may require the author to provide him with that reference.

Abstract/keywords

The abstract written must be a concise and precise summary of the work studied, highlighting the objective of the study, the materials used, the methodology employed, the results obtained, the conclusions drawn and the clinical significance of the paper if any. It must be systematically written in this order. The number of words used must not exceed the requirement for the Journal — usually not more than 150 words or 200 words.

Some Journals request for 3 to 5 keywords. Keywords must be carefully chosen to depict the exact nature of the topic researched.

Scientific Criteria

Objective(s) of study

It is very important to state the objective or objectives studied clearly and accurately. The reviewer would also evaluate whether this objective is good or adequate or whether it is inadequate.

Materials used

Merit will be given to a study using adequate materials. The study population in a clinical study must be adequately large. The number of specimens used in an animal experimentation or cadaveric study must also be adequate to be statistically significant.

A study with inadequate samples is penalised and can be rejected as unsuitable for publication by the reviewer.

Methodology employed

The correct methodology must be employed. Inappropriate or unsuitable methodology is a common cause for rejection of articles. Controls must be carefully incorporated into the study. Lack of proper controls is another common cause for failure of a particular study.

Results obtained

The results obtained from a particular research work must be thoughtfully presented in a very clear manner. Where possible, well-designed tables, appropriate figures and good illustrations would add to the quality of the work presented.

Statistical analysis must always be made from the data obtained wherever possible and the statistical methods employed clearly stated.

Results with statistical significance are more important than those where there is no statistical significance.

Conclusions drawn

It is critical that the author draws relevant, appropriate and sound conclusions based on the data obtained from the results of the study. Hypotheses are unwarranted. Extrapolations of results from animal studies to clinical applications would also be undesirable.

Clinical significance

Of course, a study with clinical significance would be rated higher than one that has little or no clinical significance.

Discussion

The reviewer would also critically analyse the quality of the discussion written by the contributor linking his work with those of important research workers on the same topic being researched upon.

The discussion should highlight important differences of his work as well as important similarities to the results of other research workers. It should emphasise new findings obtained (if any) which will make this article a significant contribution to existing literature.

Recommendation of reviewer

After performing a detailed assessment of the article based on all criteria set out — both general criteria and scientific criteria, the reviewer must decide on one of the following recommendations:

- Acceptable in present form
- Acceptable with minor revisions
- Acceptable subject to major revisions
- Unacceptable in current form but willing to reconsider if certain essential parts of the work not presented well is relooked into, the research work repeated if necessary and a new manuscript submitted within about six months.
- Totally unacceptable

Reviewer's comments

Detailed comments on all aspects of the article are required based on all the criteria the reviewer has judged. Such comments will be forwarded to the authors by the Editorial Office without disclosure of the identity of the reviewer. This is especially critical if major revision is required or the research work has been asked to be relooked into or repeated within a certain time frame.

In the latter case, once the manuscript is rewritten with major revision or resubmitted after the work is relooked into or redone, the manuscript is usually returned to the same reviewer for re-evaluation to see that he has fulfilled all the shortcomings previously noted.

Acknowledgements

The author would like to thank Dr. Wang Lihui and Ms. Jane Tan Hwee Mian for all secretarial assistance provided in the preparation of this manuscript.

Chapter 17

What Reviewers Look for in an Original Article

Joy Le Yi Wong, Wee Lin & Aziz Nather

The following is a sample marking scheme on what reviewers look for. The boxes are ranked from most to least ideal.

General Aspects

1. Abstract
   - ☐ Concise, adequate and accurate
   - ☐ Important information not included — for example, inadequate description of methodology and sample size, or poorly drawn conclusions
   - ☐ Poorly written

2. Originality
   - ☐ Novel work with new and interesting ideas
   - ☐ Provides important/useful contributions to current literature
   - ☐ Work covering an old topic

3. Clinical Impact
   - ☐ Important and practical contributions in the clinical field
   - ☐ Has less practical applications
   - ☐ Only important academically

4. Style of presentation
   - ☐ Clear and systematic
   - ☐ Difficult to follow the train of thought

Scientific Merit

1. Sample size
   - ☐ Adequate and representative
   - ☐ Inadequate or non-representative

2. Methodology and research
   - ☐ Good methodology with appropriate use of controls
   - ☐ Not clear
   - ☐ Poor methodology — vital information missing

3. Statistical analysis
   - ☐ Valid
   - ☐ Invalid

4. Discussion
   - ☐ Accurate and adequate discussion using relevant literature
   - ☐ Discussion lacks depth

5. Conclusion
   - ☐ Justified and valid
   - ☐ Invalid for reasons stated …
   - ☐ Not justified

6. Tables and figures
   - ☐ Adequate, suitable, clear and in the right quantity
   - ☐ Changes recommended
   - ☐ Unclear illustrations, insufficient illustrations or poorly designed tables

7. References
   - ☐ Accurate and adequate review of literature
   - ☐ Inadequate review of literature, does not include recent articles
   - ☐ Poor review of literature

8. Other comments for improvement

Comments to Authors

_____

_____

_____

Recommended Action

- ☐ Accept without any revision
- ☐ Accept with minor revisions
- ☐ Provisional acceptance — accept after major revision
- ☐ Reject

Chapter 18

Examining a Thesis or Dissertation

Aziz Nather

Selecting an Examiner

The Board of Examiners in Higher Degree must exercise good judgement in selecting the appropriate examiner for examining a particular thesis. The Examiner chosen must first be one who has experience in research, both clinical research and basic research. For an M.D. or D.D.S. or a Ph.D. thesis, the examiner must at least be familiar in the field of study being looked into. It would be preferable if he is an authority on that subject. He should preferably also be one who has obtained a similar degree.

In turn, the Examiner selected must decide carefully whether he would like to accept the task or not. Before making his decision, he should deliberate whether he has the necessary experience in the field investigated to examine the candidate without encountering technical difficulty in assessing the dissertation. He must also be willing to spend adequate time to examine the thesis in great detail. Examiners are required to return the evaluation within a six-month timeframe. He must be willing to accept the responsibility and complete this arduous task within this time period. If the Examiner is approached at a time when he is busy, running workshops or is going to be occupied at several conferences, in all fairness to the candidate and to the university, he should decline the invitation.

Whilst examiners are given an honorarium for the task performed, this is hardly an incentive for accepting the job. The main incentives are the prestige for being chosen to be an examiner and the satisfaction that one gets from evaluating a good piece of research. A good examiner must firstly be someone who values being called upon to accept such a job. He must be willing to accept the challenge and above all other factors be a person who enjoys doing the evaluation. An Examiner with the correct attitude will certainly be one who will do the assessment well.

General Evaluation

All Examiners must first scrutinise all available information regarding the candidate — his general background, his academic qualifications, the level of training he has mastered in orthopaedic surgery and his experience in conducting research. The Department the candidate comes from will also receive close attention — the general set up of the Department, the infrastructure and facilities available for both clinical and basic research and its academic standing for clinical and basic research. The supervisor of the candidate is also very important. Is he well qualified and well experienced to render good supervision to the candidate? The experience of the supervisor will most certainly influence the quality of the work attained by the candidate. The candidate may also have two supervisors.

Evaluating the Thesis Itself

The title of the thesis will be closely inspected. Next, the Examiner usually finds that it is very convenient and helpful to read the Summary first. This will give the examiner a panoramic view or 'bird's eye' view of the whole thesis before he starts ploughing into pages and pages of the thesis.

It is good practice for the Examiner as he reads the pages to note any comments or to make corrections on any error he comes across immediately and make a record of these comments or corrections and keep them methodically in a separate file. It is then very easy for him to refer to these relevant pages later on when he needs to compile the detailed comments for the final report.

It is also good practice not to allow more than one month to pass before starting work on the thesis. Otherwise, the Examiner is tempted to procrastinate further from month to month.

Once the examination of the thesis is started, it is also good to keep the momentum going. It is best to complete the evaluation within two to three months. He needs to spend a good one week or so to make the overall recommendation and to compile the detailed comments for the final report.

Major Considerations

In the examination of a thesis, the major considerations upon which the candidate is judged are:

- Has the candidate stated his objectives clearly?
- Has he used adequate materials to work on these objectives?
- Has he employed the correct methodology for the research work embarked upon?
- Has he derived the results properly and presented them in the best way possible? Has he analysed the results statistically?
- Has he drawn his conclusions correctly?

Additional considerations include:

- Is this piece of work original?
- Is it clinically significant?

Objectives

The objectives of the thesis must be clearly stated. They should preferably be clinically significant. It is even better if they are original. A good piece of original research can only occur if the candidate has searched the literature extensively and found a particular aspect of the subject which is clinically very important but has not been researched into.

Marks will be given for:

- Clarity of objectives stated
- Originality of work researched into
- Clinical significance of research work done

To pass a Ph.D. or an M.D. Degree or D.D.S. Degree, the candidate must produce an original piece of research. For a Master's Degree, originality is not critical though bonus points will be given for original contributions.

Materials

The materials used will be assessed as to its adequacy or inadequacy. For a clinical study, the study population should preferably be 100 or more to be clinically significant. The larger the sample size, the more significant the study.

The quality of the materials worked upon is also important. A prospective study earns more merit than a retrospective study.

Has the study population been properly standardised in terms of age, sex and race of the patients when compared with an identical group of controls? It is important that as many variables as possible are standardised for the study population, leaving only one parameter to be compared between the study group and the control group. Only then can statistical significance be obtained from the clinical study.

In the case of animal experimentation studies, the number of animals used for studying the parameter at a particular observation period should be adequate — at least four for each period of observation.

For biomechanical studies, the Examiner will also look as to whether the mechanical parameters studied such as torsion, tensile and compression strengths are done using the same specimens by employing non-destructive testing before it is loaded to failure, or whether each parameter is studied using a different cohort and employing destructive testing. The latter protocol would be rated as being clinically more significant than the former.

Methodology

Crucial to the success of the thesis is whether the candidate has used the correct methodology to perform the research. The candidate would lose valuable marks if poor methodology is employed. Of course, if he has adopted the wrong methodology he would certainly fail the examination.

For example, in an animal model studying new bone formation, more marks will be given to the candidate employing double fluorochrome labelling quantitated using undecalcified histological sections. Fewer marks will be given if the candidate employed only decalcified sections stained with H&E without utilising tetracycline labelling for quantitation.

Results

It is important to evaluate whether the candidate has interpreted his results properly. Incorrect interpretation will be disastrous. It is also important to assess whether he has presented the results in the best way possible. The data should be presented wherever possible as tables or graphs — so that they could be more easily understood. Statistical analyses should be performed and the statistical methods employed clearly described.

Furthermore, if the results were shown to be statistically significant, the thesis would be considered to be more successful than if the results were not statistically significant.

Conclusions

The conclusions drawn by the candidate are also important. Conclusions should be correctly made based on the results obtained from the research work. The candidate would be penalised if the conclusions drawn were inappropriate or far-fetched.

Extrapolations should be avoided. For example, conclusions made from an experimental surgery research involving rabbits should not be extrapolated to be applicable clinically to man.

There is certainly no place to make any hypothesis. Unwarranted recommendations should not be made. To be on the safe side, tt is better not to make any recommendation at all from the research work than to make one which is found to be unwarranted.

It does not matter if the conclusions reached are not what the candidate and the supervisor expected. So long as the objectives have clearly defined, the materials used were adequate, the methodology employed were correct, the results analysed correctly and the conclusions correctly derived, the candidate has fulfilled all the requirements of a thesis and should be deemed to have passed the examination. It does not matter whether the conclusions were different from what even the Examiners expected.

Other Considerations

Other considerations assessed by the Examiner include:

- Is the thesis well written?
- Is the thesis well illustrated?

A thesis well written with good English and a thesis well illustrated with good photographs and clearly designed tables will of course also be well rewarded by the Examiner.

Final Consideration

- Has this piece of work good potential?

Finally, the Examiner will also evaluate the potential of this thesis to lead to publications. Bonus points would be awarded if the Examiner feels that the research work could lead to one or more publications in top-ranked international refereed Journals.

Writing an Examiner's Report

The Examiner's report to be given to the Board of Examiners in Higher Degrees should include:

- Overall recommendation on the thesis
- Detailed comments on the thesis

The comments should be fairly detailed; the amount of details to be provided depends on the particular merits or faults of the thesis/dissertation to which the Examiner wishes to draw attention to. The report should on the average range in length between two and six A4 pages typed in double spacing. In cases where revision is recommended, a more detailed report would be of considerable assistance to the candidate and to his supervisor.

Standards Adopted for Examination

Although no specific standards have been prescribed for Master's and doctoral theses/dissertations, the standards to be accepted should comparable to the standards expected by reputable universities in the UK or USA.

It is mandatory that the Examiner must not under any circumstances, communicate with the candidate, the candidate's supervisor and the other Examiners on matters relevant to the thesis/dissertation examination.

Master's Degrees

The Examiner may at his own discretion set such further tests as he thinks fit to enable him to assess the thesis or dissertation. Master's candidates are not required to be orally examined. The Examiner may recommend to the Board of Examiners that the candidate be passed or failed or referred for further work. In the latter case, the thesis must be resubmitted for further consideration.

Degree of Ph.D./M.D./D.D.S

For a Ph.D. (Doctor of Philosophy), the thesis must contain original work or critical interpretation worthy of publication.

For an M.D. (Doctor of Medicine), the thesis must be an original piece of work on a subject in the field of clinical research in a branch of medicine.

For a D.D.S. (Doctor of Dental Surgery) the thesis should make an original and substantial contribution to a branch of dentistry.

The candidate for any of these degrees must be examined orally. The Panel of Oral Examiners comprises only internal members in the university the candidate is registered with. The external examiner is not required to be present during the oral examination. The question/issues raised by the external Examiner in the detailed comments of his report will be put forward to the candidate by the Oral Panel.

Examiner's Overall Recommendation

After examination of the thesis, the Examiner can recommend that the candidate be

- Awarded the degree without further examination — unconditionally;
- Awarded the degree subject to the candidate making minor/technical amendments/corrections. Such minor amendments include spelling and typing errors, use of terminologies, language and other changes deemed to be minor or technical in nature;
- Awarded the degree subject to the candidate making amendments of a reasonably substantial nature. Such substantial amendments include rewriting a section of the dissertation, providing additional

clarification, redoing of tables and graphs and other changes which did not affect the overall quality of the thesis;
- Not awarded the degree but be permitted to resubmit the thesis in a revised form after a further period of study and research; or
- Not awarded the degree.

Examiner's Detailed Comments

The Examiner must state concisely the grounds on which he bases the overall recommendation indicating where appropriate the strengths and weaknesses of the thesis.

Such detailed comments could be made in two parts. The first part contains comments on the thesis which are confidential for transmission to the Board of Examiners in Higher Degrees. The second part contains comments on the thesis, and where applicable, list of minor corrections/typographical changes/amendments/revisions to be made in the thesis for transmission to the candidate.

In recording the detailed comments, the Examiner would deliberate on the following aspects of the thesis:

- Summary/abstract
- Review of literature
- Objectives
- Materials used
- Methodology employed
- Results obtained
- Conclusions drawn
- Quality of discussion
- Bibliography

Assessment Format

In some universities, the Examination Board has already considered the various factors to be assessed and requires the Examiner to evaluate these factors based on given weighted scores it has decided upon.

An example of such an assessment format used by one university is as follows:

	Weightage (%)
Presentation	15
Abstract	10
Introduction and review of literature	10
Objectives	5
Methodology	15
Results	15
Discussion	25
Bibliography	5
Total	100

Acknowledgements

The authors should record their appreciations to various individuals giving help or support in completing the thesis e.g.

— for secretarial assistance
— for statistical analysis etc.

Chapter 19

What Reviewers Look for in Writing a Thesis

Zest Yi Yen Ang, Joy Le Yi Wong & Aziz Nather

Summary

☐ **Tier 1**

- ✓ Provides bird's eye overview of entire thesis
- ✓ Easily understood at a glance

☐ **Tier 2**

- ✓ Not comprehensive
- ✓ Does not provide full picture of thesis

☐ **Tier 3**

- ✓ Poorly written with errors

Thesis Statement

☐ **Tier 1**

- ✓ Contestable stand
- ✓ Clear and specific

- ✓ Makes significant links to larger implications
- ✓ Provides new insights

☐ **Tier 2**

- ✓ General
- ✓ Based on old topics
- ✓ No new insights
- ✓ Not supported by essay

Objectives

☐ **Tier 1**

- ✓ Clearly stated
- ✓ Clinically significant
- ✓ Original

☐ **Tier 2**

- ✓ Lacks originality
- ✓ But makes important/useful contributions to current literature

☐ **Tier 3**

- ✓ Vague
- ✓ No clinical significance

Methodology

Overall

☐ **Tier 1**

- ✓ Correct methodology employed to perform research
- ✓ Prospective rather than retrospective study
- ✓ Clearly documented

- [ ] **Tier 2**

    ✓ Poor methodology employed to perform research
    ✓ Unclear documentation

- [ ] **Tier 3**

    ✓ Incorrect methodology
    ✓ Poor documentation with missing vital information

Study Population

- [ ] **Tier 1**

    ✓ Appropriate and representative population size
    ✓ Standardised in terms of age, sex and race
    ✓ Correct use of controls

- [ ] **Tier 2**

    ✓ Inadequate and non-representative population size
    ✓ Wrong use of controls

Results

- [ ] **Tier 1**

    ✓ Correct interpretation and analysis of data
    ✓ Clear, effective presentation
    ✓ Good use of tables and graphs
    ✓ Valid statistical analysis of results performed with clear description of tests
    ✓ Results proven to be statistically significant

- [ ] **Tier 2**

    ✓ Changes recommended to improve figures, tables or graphs
    ✓ Statistical tests can be better described

☐ **Tier 3**

- ✓ Incorrect interpretation and invalid analysis of data
- ✓ Unclear illustrations, poorly designed tables and figures
- ✓ Results statistically insignificant

Discussion

☐ **Tier 1**

- ✓ Specific and focused
- ✓ Displays strong understanding and command of relevant literature
- ✓ Anticipates and refutes counter-arguments
- ✓ Displays a groundbreaking level of thought

☐ **Tier 2**

- ✓ Lacks depth
- ✓ Poor review of relevant literature

Conclusion

☐ **Tier 1**

- ✓ Correct, logical conclusion drawn based on findings
- ✓ Substantiated recommendations

☐ **Tier 2**

- ✓ Invalid due to the following reasons: _____

☐ **Tier 3**

- ✓ Far-fetched
- ✓ Unsubstantiated claims

Overall

- ✓ Error free
- ✓ Original
- ✓ Good potential for publication in international journals
- ✓ Good grammar
- ✓ Avoids sweeping generalisations; no words such as 'all', 'none' or 'every'
- ✓ Clear and systematic presentation
- ✓ Well-illustrated
    - ⇨ Good photographs
    - ⇨ Clearly designed tables

Overall Recommended Action

- ➤ Accept without any revision
- ➤ Accept with minor revisions
- ➤ Provisional acceptance — accept after major revision
- ➤ Reject

Comments to Authors

Based on merits/faults of particular thesis examiner believes should be focused on. It is split into two parts:

1) Comments on the thesis which are **confidential** for transmission to the **Board of Examiners** in Higher Degrees.
2) Comments on the thesis, and where applicable, list of minor corrections/typographical changes/amendments/revisions to be made in the thesis for transmission to the **candidate**.

Chapter 20

Objective Evaluation of Research Output

Aziz Nather

Introduction

It is difficult to objectively evaluate the research output of an individual or an institution. Very often, too much weightage is given to subjective assessments by department heads or peers. The following criteria should be used to achieve a more objective assessment:

1. Publications
2. Research grants obtained
3. National/international research awards received
4. Membership on National Research Committees/Editorial Boards
5. Peer review

Publications

The most objective parameter available to evaluate the research output of an individual is to scrutinise closely the list of publications the individual has published. Publications include:

- Peer-reviewed, original research articles
- Non-peer-reviewed articles
- Book chapters/review articles

Journals in which articles are published are ranked and weightage is given to publications in highly ranked journals. The number of citations of a researcher's work by peers is another assessment of the quality of the research product. For these reasons, publications provide the main avenue by which the research work done by an academic can be objectively evaluated.

Ranking of Journals

Research work published in a refereed or peer-reviewed scientific journal is ranked as being of higher quality than one published in a non-refereed or non-peer-reviewed journal.

Peer-Reviewed Original Research Articles

In the ranking of journals, there are three important criteria[1]:

- Absolute citation frequency
- Immediacy index
- Impact factor

The **absolute citation frequency** refers to the number of times a journal is cited in other journals. The disadvantage of using this criterion is that a journal published frequently will theoretically have an advantage over one appearing less frequently.

The **immediacy index** is a measure of how quickly the average article in a particular journal is cited, e.g., the 2000 immediacy index of Journal A is obtained by dividing the number of all 2000 journal citations of Journal A articles published in 2000, by the total number of articles published by Journal A in 2000. Again, a journal published more frequently, e.g., monthly will have an advantage over a journal published quarterly or semi-annually.

The **impact factor** is a measure of the frequency with which an average article in the journal is being cited in a particular year. Thus, the 1999 impact factor of Journal B is calculated by dividing the number of all citations of articles in 1999, that Journal B published in 1997 and 1998, by the total numbers of articles Journal B published in these two years. The impact factor is by far the most reliable index in the ranking of journals. It is not affected by the size of journal (large versus small), the frequency of publication of the journal (more frequent versus less frequent) and the age of the journal (older versus newer). Impact factors are published in annual volumes of the Journal Citation Reports and in various citations indices published by the Institute for Scientific Information.

Non-Peer-Reviewed Articles

Whilst the number and significance of original research articles in peer-reviewed journals is the most important part in the evaluation of an academic's list of publications, particularly in journals ranked as A* and A, publications in non-peer-reviewed journals should not be dismissed as not important at all. The number of such articles published are also important. These articles reflect the intellectual and scholarly activity of the individual.[2]

Book Chapters/Review Articles

It is also important to take into consideration the total number of book chapters or review articles published by the researcher. This is because book chapters or review articles reflect the scholarship of integration, another component of the intellectual activity of the individual.[2]

Theses/Monographs

In the assessment of research output research output of an individual, points should also be awarded for a M.D., M.S. or Ph.D. thesis and for monographs or entire books written. Book editorship without any chapter being written by the editor will receive fewer points than when the book edited has several chapters contributed by the researcher. More points should be awarded when the entire book is written by a sole author.

Research Grants Obtained

In the assessment of research output, points should also be awarded to the number of research grants obtained by the research worker. More points should be given for grants given for "mega-projects" and for grants awarded by the National Medical Research Council (NMRC). Points should also be given to grants obtained from other prestigious sources e.g., Agency for Science, Technology and Research (A*STAR), Economic Development Board (EDB) and Totaliser Board. The number of prestigious grants awarded reflects good peer acceptance of the individual's ability to conduct research projects.

National/International Research Awards Received

Merit points should be given to recipients of international research awards, e.g., the Volvo Award for spine basic research work and the Young Investigator Award for International Conferences (e.g., Orthopaedic Research Society Meeting).

Points should also be awarded to recipients of national research awards, e.g., Young Investigator Award from A*STAR and other national conferences such as the Singapore Orthopaedic Association Meeting.

Membership on National Research Committees/Editorial Boards

The research standing of an individual as recognised by his peers is also reflected by his invitation to serve on national research committees, e.g., NMRC, Faculty Research Committee, National Advisory Bioethics Committee, etc.

Likewise, points should be awarded to individuals invited to serve in Editorial Boards of peer-reviewed journals, e.g., *Journal of Orthopaedic Research*, *Journal of Musculo-Skeletal Research*, etc.

Peer Review

The final measure of the research standing of an academic is rated by peers both within the faculty and outside the institution.[3–5]

Peer review is also responsible not only for ranking of publications, but also for the number of research grants obtained by an individual, the number of research awards given to a research worker internationally and nationally, and also the number of Research Committees and Editorial Boards a research worker has been invited to serve on.

References

1. Garfield, E., Malin, M. V. & Small, H. (1978). Citation data as science indicators. In: Elkana, Y., Lederberg, J., Merton, R. K., Thackray, A. & Zuckerman, H. (eds.). *Toward a Metric of Science: The Advent of Science Indicators*. New York: John Wiley & Sons, pp. 179–207.
2. Holmes, E. W., Burks, T. F., Dzau, V., Hindery, M. A., Jones, R. F., Kaye, C. I., Korn, D., Limbird, L. E., Marchase, R. B., Perlmutter, R., Sanfilippo, F. & Strom, B. L. (2000). Measuring contributions to the research mission of medical schools. *Acad. Med.* **75**: 304–313.
3. Cresswell, J. W. (1985). Measure of faculty research performance in Faculty Research Performance, Lessons from the Science and Social Sciences. Report 4, *Ashe-Eric Higher Education Reports*, pp. 7–14.
4. Donald, J. G. (1984). Quality indices for faculty evaluation. *Assess. Eval. High. Edu.* **9**: 41–52.
5. Kumar, V. P., Satku, K., Wee, J. T. K. & Pho, R. W. H. (1992). Objective Assessment of Research Performance. *Ann. Acad. Med.* **21**, 436–438.

Appendix

Glucon

Supports Healthy Joints

Crystalline Glucosamine Sulphate 628 mg
+
Chondroitin Sulphate 400 mg

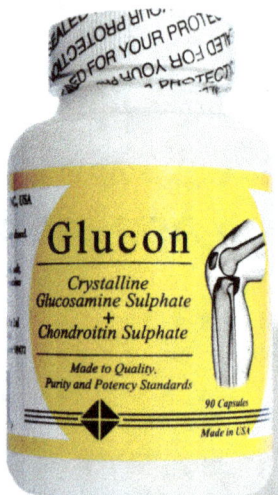

Made in USA

 MD PHARMACEUTICALS

MD Pharmaceuticals Pte Ltd
896 Dunearn Road, #02-01A Sime Darby Centre
Singapore 589472. Phone:64654321

MD₃

Vitamin D3 1000 IU

Supports bone health and offers various health benefits

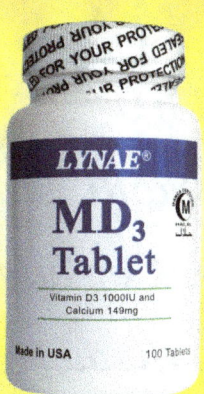

Made in USA

MD Pharmaceuticals Pte Ltd
896 Dunearn Road, #02-01A Sime Darby Centre
Singapore 589472. Phone:64654321

Pro-Gut™

**Promotes a healthy balance of digestive flora
Helps reinforce intestinal functions...naturally**

- *9 Billion CFU*
- *8 Live Probiotic Strains*
- *Vegetarian Capsules*
- *Made in Canada*

MD Pharmaceuticals Pte Ltd, 896 Dunearn Road, #02-01A Sime Darby Centre, Singapore 589472. Phone:64654321

For Maintenance of Healthy Joints

GS1

Glucosamine Sulphate 2KCl
1500 mg Tablets

GLUCO-S1500

Glucosamine Sulphate 1500 mg Sachets

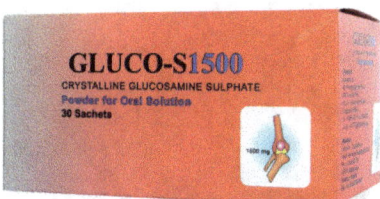

MD Pharmaceuticals Pte Ltd
896 Dunearn Road, #02-01A Sime Darby Centre
Singapore 589472. Phone:64654321

Everything you love about foam dressings *and more*

- Protective top layer
- Soft FOAM pad
- AQUACEL™ layer
- Gentle silicone border

Now only one dressing offers the comfort and simplicity of FOAM plus the **healing benefits of an AQUACEL™ layer**

- Gentle silicone border designed to adhere to surrounding skin, not the wound bed
- Available in a range of adhesive and non-adhesive sizes

AQUACEL™ Foam

To find out more about AQUACEL™ foam dressing or to arrange a visit from your ConvaTec representative, please call 65-6245-9838.

www.convatec.com/aquacelfoam

AQUACEL, the AQUACEL logo, ConvaTec, the ConvaTec logo, Hydrofiber and the Hydrofiber logo are trademarks of ConvaTec Inc., and are registered trademarks in the U.S.
© 2012 ConvaTec Inc.

AP-011757-MM

Introducing AQUACEL™ Ag Foam.

- **Killed more *P. aeruginosa* and *S. aureus* bacteria** beneath the dressing than other silver foam dressings tested*[1]

- **The only silver foam dressings with Hydrofiber™ Technology that micro-contour** to the wound bed, minimizing dead space where bacteria can grow*[2]

- **Helps to reduce** the risk of maceration*[3,4]

- Available in a wide range of shapes and sizes

65-6245-9838 / AQUACEL.com/foam
*As demonstrated *in vitro*

1. The antimicrobial activity of AQUACEL™ Ag Foam adhesive using a simulated shallow wound microbial model. *Microbiological Application*. WHRI3771 MA221. 2013. Data on file, ConvaTec. 2. *In vitro* testing of AQUACEL™ Ag Foam and Competitor Dressings – Intimate Contact. Market Support. WHRI3661 MS100. 2013. Data on file, ConvaTec. 3. Waring MJ, Parsons D. Physico-chemical characterisation of carboxymethylated spun cellulose fibres. *Biomaterials*. 2001;22(8):903-912. 4. Cook L, Baker C. AQUACEL™ Foam dressing: A case study demonstrating its effectiveness in managing the complications of wound exudate under compression bandaging. Poster presented at: Wounds UK Conference. November 12-14, 2012, Harrogate.
®/™ AQUACEL and Hydrofiber are trademarks of ConvaTec Inc. All other trademarks are property of their respective owners. ©2013 ConvaTec Inc. AP-013501-MM

Wound healing has always had villains.
Now it has a hero.

EXUDATE **INFECTION** **BIOFILM**

Our breakthrough new dressing attacks the key local barriers to wound healing, even biofilm.

TWO POWERFUL TECHNOLOGIES

Our proven **Hydrofiber™ Technology** absorbs and retains excess exudate to help create an ideal healing environment.*[1-5] And now our revolutionary new **Ag+ Technology** destroys biofilm and kills infection-causing bacteria.*[6-8]

See how it helps you save the day at **www.convaTec.com**.

*No dressing does more.^*

65-6245-9838 / www.convaTec.com

*As demonstrated in vitro.
^Defined as the ability to manage exudate, infection and biofilm, as demonstrated in vitro.
1. Newman GR, Walker M, Hobot JA, Bowler PG. 2003. Scanning electron microscopic examination of bacterial immobilisation in a carboxymethyl cellulose (AQUACEL™) and alginate dressings. Biomaterials; 24: 853–860. 2. Bowler PG, Jones SA, Davies BJ, Coyle E. 1999. Infection control properties of some wound dressings. J. Wound Care; 8: 499-502. 3. Walker M, Bowler PG. Cochrane CA. 2007. In vitro studies to show sequestration of matrix metalloproteinases by silver-containing wound care products. Ostomy/Wound Management. 2007; 53: 18-25. 4. Assessment of the in vitro Physical Properties of AQUACEL EXTRA, AQUACEL Ag EXTRA and AQUACEL Ag+ EXTRA dressings. Scientific background report. WHRI3917 TA297. 2013. Data on file. ConvaTec Inc. 5. Physical Disruption of Biofilm by AQUACEL® Ag+ Wound Dressing, Scientific Background Report, WHRI3850 MA232, 2013, Data on file, ConvaTec Inc. 7. Antimicrobial activity and prevention of biofilm reformation by AQUACEL® Ag+ EXTRA dressing. Scientific Background Report. WHRI3875 MA239. 2013. Data on file, ConvaTec Inc. 8. Antimicrobial activity against CA-MRSA and prevention of biofilm reformation by AQUACEL® Ag+ EXTRA dressing. Scientific Background Report. WHRI3875 MA239. 2013. Data on file. ConvaTec Inc.

®/™ Indicates trademarks of ConvaTec Inc. ©2013 ConvaTec Inc. AP-014181-MM

REDUCED INFECTION BY 67%[1a,2b]
REDUCED BLISTERING BY 88%[1a,2b]

Are these figures you can operate with?

AQUACEL™ Ag SURGICAL cover dressing provides a waterproof viral and bacterial barrier* and utilizes patented Hydrofiber® Technology with ionic silver to manage exudate and provide antimicrobial activity against a variety of pathogens, including MRSA and VRE, as demonstrated in vitro.[3]

AQUACEL™ Ag SURGICAL cover dressing is supported by evidence showing significant reduction in the incidence of superficial surgical site infection, blistering and delayed discharge as compared to a non-woven post-operative surgical dressing regimen.[1a, 2b] Make it part of your protocol of care and see the difference it may potentially make in infection control and positive clinical outcomes.

AQUACEL™ Ag SURGICAL. The seal of excellence

 Find out more about AQUACEL™ Ag SURGICAL cover dressing, visit www.convaTec.com or call 65-6245-9838

*When intact and there is no leakage

1. Clarke JV, Deakin AH, Dillon JM, Emmerson S, Kinninmonth AWG. A prospective clinical audit of a new dressing design for lower limb arthroplasty wounds. J Wound Care 2009;18(1):5-11 2. Laboratory Test Comparison of AQUACEL® Surgical Dressing 'New Design' and the Jubilee Method of Dressing Surgical Wounds. WHRI0264.TA130, October 7, 2009. Data on file, ConvaTec. 3. Jones SA, Bowler PG, Walker M, parsons D. Controlling wound bioburden with a novel silver-containing Hydrofiber dressing. Wound Repair Regen. 2004;12(3):288-294.

®/™ indicates trademarks of ConvaTec Inc, unless otherwise noted. AQUACEL, DuoDERM and Hydrofiber are registered trademarks in the United States. Mepore is a trademark of Mölnlycke Health Care. AP-008415-MM/US/CA. © 2009 ConvaTec Inc.

Acelity™

You already know our products. **Now meet Acelity.**

We offer a portfolio of advanced wound healing products
and support to better meet your needs.

Our commitment is to deliver proven solutions and specialised
knowledge to help people heal and be whole again.

VeraFlo™ Therapy
using V.A.C.Ulta™ Negative Pressure
Wound Therapy System

CelluTome™
Epidermal Harvesting
System

Prevena™
Incision Management
System

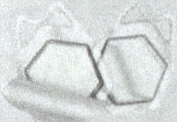
PROMOGRAN®
Collagen/ORC Dressing
PROMOGRAN PRISMA®
Collagen/ORC/Silver Dressing

**SILVERCEL®
NON-ADHERENT**
Antimicrobial
Dressing

INADINE®
Povidone Iodine
Non-Adherent Dressing

**ACTISORB®
PLUS 25**
Activated Charcoal
Dressing with Silver

To learn more, please visit acelity.com

 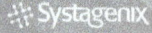

Acelity Companies

AUTOMATED EPIDERMAL HARVESTING

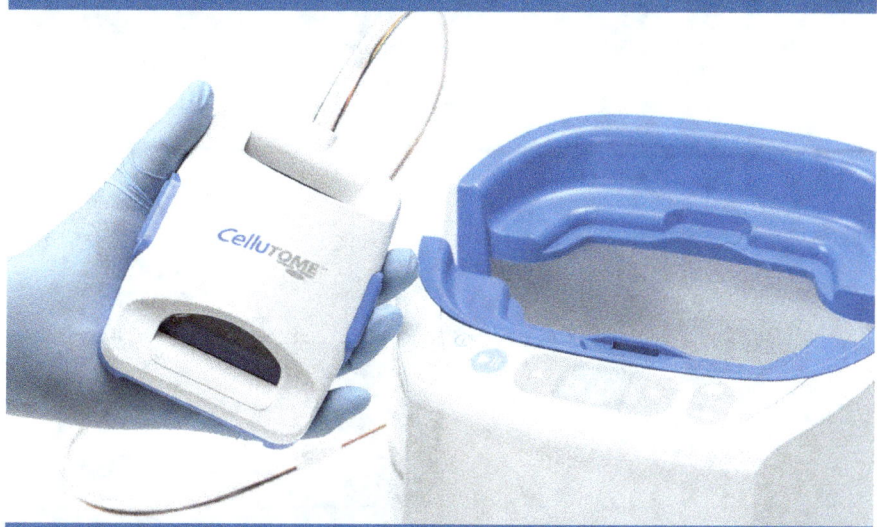

When there's a need for epidermis, **PRESS START**

The **Cellutome**™ Epidermal Harvesting System is an innovative epidermal harvesting device which can be easily integrated into your wound care clinic

- Automated, precise and reproducible process
- Fast harvesting procedure takes, on average, 45 minutes
- Comprehensive training from a KCI representative in less than one hour

NOTE: Specific indications, contraindications, warnings, precautions and safety information exist for KCI Products and therapies. Please consult a physician and product instructions for use prior to application. This material is intended for healthcare professionals.

©2014 KCI Licensing, Inc. All rights reserved. All other trademarks designated herein are proprietary to KCI Licensing, Inc., its affiliates and/or licensors.
DSL#14-0473.CT.SG (Rev. 6/15)

ACTIVE INCISION MANAGEMENT

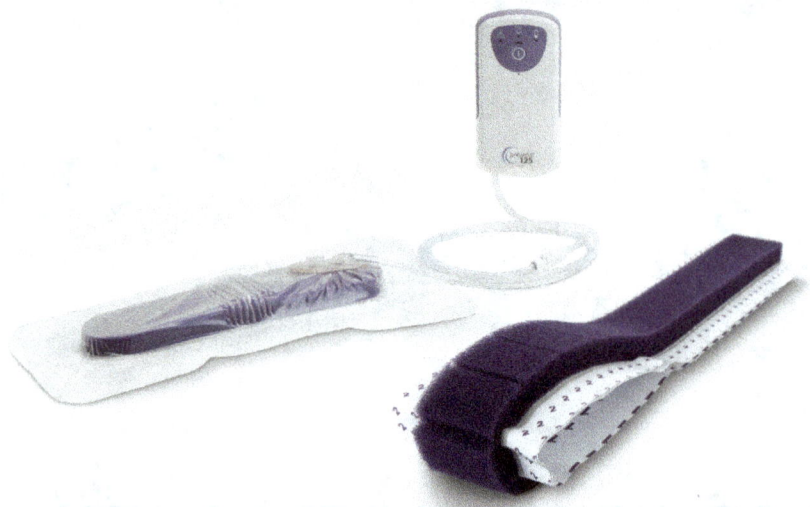

Prevena™ Therapy manages and protects closed surgical incisions

Prevena™ Therapy: a unique design for incision management

- Helps hold incision edges together
- Removes exudate
- Acts as a barrier to external contamination
- Delivers continuous negative pressure at -125mm

NOTE: Specific indications, contraindication, warnings, precautions and safety information exist for KCI Products and therapies. Please consult a physician and product instructions for use prior to application. This material is intended for healthcare professionals.

KCI
An Acelity Company

©2014 KCI Licensing, Inc. All rights reserved. All trademarks designated herein are proprietary to KCI Licensing, Inc., its affiliates and/or licensors. DSL#14-0473.P.SG (Rev. 7/15)

Acelity

v.a.c.® therapy

BE CONFIDENT IN YOUR NPWT CHOICE

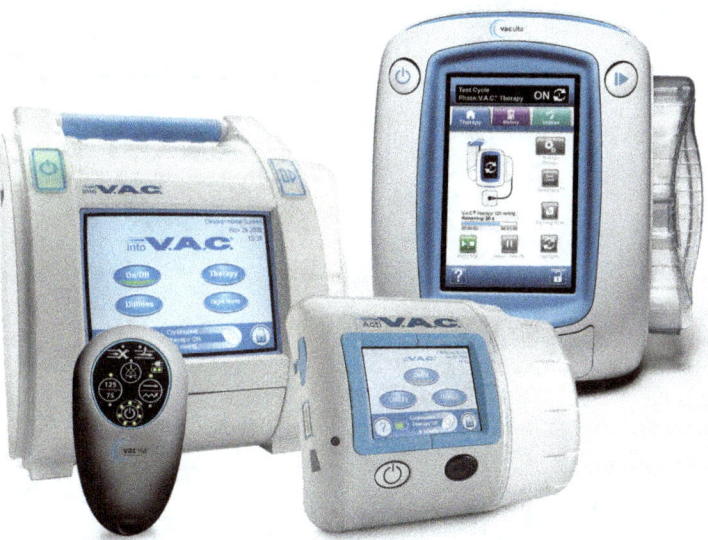

V.A.C.® Therapy with SensaT.R.A.C.™ Technology

Ask us about:

- SensaT.R.A.C.™ Technology
- Most published NPWT clinical evidence

NOTE: Specific indications, contraindication, warnings, precautions and safety information exist for KCI Products and therapies. Please consult a physician and product instructions for use prior to application. This material is intended for healthcare professionals.

An Acelity Company

©2014 KCI Licensing, Inc. All rights reserved. All trademarks designated herein are proprietary to KCI Licensing, Inc., its affiliates and/or licensors. DSL#14-0473.VAC.SG (Rev. 7/15)

Diabetic foot ulcer

smith&nephew

Diabetes is a leading cause of lower limb amputation and 25% of people living with diabetes will develop diabetic foot ulcer. However up to 85% of DFU-related amputations can be avoided when an effective care plan is adopted.[1]

Choosing the right management choice can improve the outcome of diabetic foot ulcer.

INTRASITE° GEL
Hydrogel Wound Dressing

ACTICOAT° Flex
Nanocrystalline Silver† Dressing

ALLEVYN°
Ag Gentle Border
Hydrocellular Dressing

DURAFIBER°
Gelling Fiber Dressing

IODOSORB°
Ointment
Cadexomer Matrix with Iodine

Reference
1 International Best Practices Guidelines: Wound management in diabetic foot ulcers. Wounds International, 2013. Available from: www.woundsinternational.com.

† Nanocrystalline silver is a patented technology of NUCRYST Pharmaceuticals Corp.

Smith & Nephew Pte Ltd, 1A International Business Park #09-03 Tolaram, Singapore 609934
Telephone: +65 6270 0552, Fax: +65 6272 6698, Email: ASEAN@smith-nephew.com

www.smith-nephew.com

*Trademark of Smith & Nephew. Registered in US Patent and Trademark Office.
2014 2904 SIN/WM/ DFU ADTEMPLATE/01/01

Pressure Ulcer

Pressure ulcers remain one of the major healthcare problems around the world with over 65,000 patients dying from related complications yearly.[1, 2]

With proper prevention and pressure ulcer maintenance measures, minimising pressure ulcer related incidents is achievable.

DERMAPAD*
Polymer Gel

ALLEVYN* Ag Gentle Border
Hydrocellular Dressing

INTRASITE*
Hydrogel Wound Dressing

IODOSORB* Ointment
Cadexomer Iodine

RENASYS*
Negative Pressure Wound Therapy

Reference
1. Pressure Ulcers: A patient safety issue. National Center for Biotechnology Information. Retrieved from www.ncbi.nlm.nih.gov/books/NBK2650/. Accessed 01 December 2013.
2. Wellstar International Inc., 14 January 2009. Marketwired. Wellstar/TMI Announces Pending Pressure Ulcer Study on Leg and Foot Ulcers. Retrieved from www.marketwired.com/printer_friendly?id=1246236.

Smith & Nephew Pte Ltd, 1A International Business Park #09-03 Tolaram, Singapore 609933
Telephone: +65 6270 0552, Fax: +65 6272 6698, Email: ASEAN@smith-nephew.com

* Trademark of Smith & Nephew. All trademarks acknowledged.

www.smith-nephew.com

2014 2904 SIN/WM/PU ADTEMPLATE/01/01

smith&nephew

Protected: Because every break in the skin is at risk of infection

Improve the outcome of non-surgical wound healing with Smith & Nephew dressing solution.

PRIMAPORE®
Adhesive Wound Dressing

OPSITE® Post-Op Visible
See-through Absorbent Dressing with Waterproof and Bacterial-proof Film

PICO®
Single Use Negative Pressure Wound Therapy

ACTICOAT® Flex
Nanocrystalline Silver Dressing

Take control of Surgical Site Infection. Visit www.smith-nephew.com for more information.

Smith & Nephew Pte Ltd. 1A International Business Park #09-03 Tolaram Singapore 609933 T: +65 6270 0552 F: +65 6272 6698 E: ASEAN@smith-nephew.com
™Trademark of Smith & Nephew. Registered in US Patent and Trademark Office.
2014 2504 SIN/WM/SSI AD TEMPLATE/01/01

TIME[1] to prepare wound bed for optimal healing.

Developed by the International Wound Bed Preparation Advisory Board, TIME framework aims to identify the barriers to healing and recommend a plan of care to remove the barriers to wound healing. Thus reducing overall financial burden and treatment time to the patients.

Goal: Assess wound characteristics and determine the type of wound bed preparation goal that is essential to optimize wound healing.

	T Tissue Non-Viable	**I** Infection & Inflammation	**M** Moisture Balance	**E** Edge of wound
Wound Factors	 Necrotic Tissue or Slough Present	 Increased Exudate, Surface Discolouration or Increased Odour	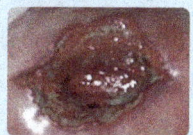 Risk of Maceration (heavy exudate) or Dessication (dry wound bed)	Chronic Wound With Prolonged Inflammation
Clinical Actions	**Debridement** Necessary to remove defective tissues, slough, exudate and debris that are known to delay healing and cause infection build-up.	**Remove or Reduce Bacterial Load** Consider biofilms in chronic wounds where protective layers of built-up become resistant to the action of antimicrobials, including antibiotics.	**Restore Moisture Balance** Essential for wound healing to be achieved.	**Address T/I/M Issues** If the wound edges failed to contract and reduce in size, the edge of wound will not epithelialize unless the wound bed is well prepared.
	Types of Debridement	**Types of Antimicrobial**	**Types of Exudate Management**	**Wound Closure**
Product Solutions	INTRASITE° Gel Hydrogel Wound Dressing  — Autolytic	ACTICOAT° Nanocrystalline Silver Dressing  — Nanocrystalline Silver *Shown in-vitro to prevent the formation of biofilms.	ACTICOAT° Absorbent Nanocrystalline Silver Dressing — Alginate	If edge of wound does not advance after 2-4 weeks, reassess intervention.
	IODOSORB° Cadexomer Carrier with 0.9% Iodine — Autolytic	ALLEVYN° Ag Hydrocellular Antimicrobial Dressing — Silver Sulfadiazine	ALLEVYN° Hydrocellular Dressing — Hydrocellular	PICO° Single Use Negative Pressure Wound Therapy — Negative Pressure Wound Therapy
	IRUXOL° Mono Collagenase Ointment 1.2U/g — Enzymatic	IODOSORB° Cadexomer Carrier with 0.9% Iodine — Cadexomer Iodine *Shown to disrupt and prevent the formation of biofilms.	IODOSORB° Cadexomer Carrier with 0.9% Iodine — Cadexomer Iodine	
	VERSAJET° Hydrosurgery system  — Hydrosurgical	ALGISITE° Ag Calcium Alginate with Silver  — Silver Alginate	RENASYS° Negative Pressure Wound Therapy — Negative Pressure Wound Therapy	
Wound Outcomes	Viable (vascularized) Wound Bed 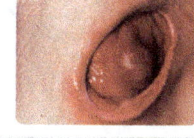	Reduced Inflammation 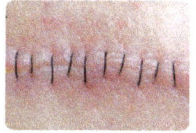	Optimal Moisture Balance 	Epithelialized Wound 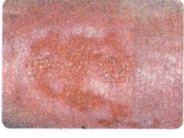

www.ingramcontent.com/pod-product-compliance
Lightning Source LLC
Chambersburg PA
CBHW070307230426
43664CB00015B/2658